Contradictions
Neuroscience and Religion

José M. Musacchio

Contradictions

Neuroscience and Religion

 Springer

José M. Musacchio
Department of Pharmacology
New York University Langone Medical Center
New York, NY 10016
USA

SPRINGER-PRAXIS BOOKS IN POPULAR SCIENCE
SUBJECT *ADVISORY EDITOR*: Stephen webb, B.sc, ph,D.

ISBN 978-3-642-27197-7   ISBN 978-3-642-27198-4 (eBook)
DOI 10.1007/978-3-642-27198-4
Springer Heidelberg New York Dordrecht London

Library of Congress Control Number: 2012938654

Printed on acid-free paper

Springer is part of Springer Science+Business Media (www.springer.com)

*To my daughters, Michèle Musacchio
and Andrea Barnhill*

Preface

The aim of *"Contradictions"* is to explore the problems that are implicit in the simultaneous and naive acceptance of science, religion, and abstract objects that permeate our culture. Scientific and empirical knowledge are coherent, but they seem to be at odds with popular religious and spiritual beliefs which contradict not only each other, but also the unity of truth. *"Contradictions"* has a long history that is deeply ingrained in my life and work. My brother Willy died when I was 12 years old, so I became preoccupied with the problem of survival after death. I was very sad, but everybody told me that we have a body and a soul and that his soul survived his death. I had no doubt that we have a physical body, and religions and the subjective view tell us that we also have a "spiritual" component that seems nonphysical. I suspected, however, that faith was not a rigorous method for answering fundamental questions, because it gave rise to innumerable contradictory religions that have created animosities and caused many wars in Europe and elsewhere.

I gradually concluded that the shortest route to learn about human nature was to become a physician, like my father and grandfather. In medical school, I was captivated by most subjects, but mainly by the brain and its diseases. I was initially fascinated and later terrified by the disintegration of the mind that I witnessed in the neurology wards during my student days in La Plata, Argentina, and later during my residency training at Bellevue Hospital in New York City. By then, I thought that the physical nature of the mind was well established, because the agents that modify it, such as words, psychoactive drugs, diseases, and trauma were all physical.

There are of course additional factors that are easily acceptable, but confuse the picture. For example, the external world is concrete, but what we call *abstract objects* such as those of ethics, esthetics, and mathematics are generally considered nonphysical entities. Since Plato, pure forms have been conceived as immaterial or abstract. The problem with abstract concepts, pure forms, and numbers is that brain diseases can wipe out the information about words, facts, faces, or even half of our bodies in a *modular fashion*, as if this information were stored like the individual files in a computer. Additionally, we have all seen people with Alzheimer's disease, whose minds progressively disintegrate to the point that they do not even recognize their close relatives. My conclusion about the physical nature of the mind is not unique, but implies that our *soul* is also physical, an opinion that many authors of popular books are reluctant to express.

Since my student days I have been interested in the philosophy of mind, but I thought that it should be complemented with what we know from neuroscience. In the process, I realized that there is something very peculiar about our brain, namely, that *we cannot sense it*, because sensing the sensors that sense the sensors and so on, would create a never-ending chain of events, which philosophers call *infinite regress*. Thus, I discovered that the imperceptibility of the brain was logically necessary, but leads to the universal but deeply misleading intuition that we have a *supernatural soul*. This is of course an attractive idea, because a supernatural soul should be immortal. Unfortunately, neuroscience indicates otherwise.

The problems of human nature are complex and intertwined, so my book is not only about what we are, but also about the nature of our knowledge. Inspired by reading C. P. Snow's *"The Two Cultures"*, [1] I realized that the arbitrary division of our knowledge into disciplines allows us to create tight compartments to protect incongruent beliefs. For example, we could justify almost any belief by taking refuge in a particular religion, philosophical doctrine, or narrow academic discipline. The trick is to convince other scholars that other disciplines have no relevance to the subject or that they have an "independent nature". We tend to forget that the *unity of truth and the universe* supersedes our provincial cultures. Opinions are subjective, but truth is universal, and it has a unity and beauty that leave no room for contradictions. Today, the unity of knowledge and truth are both essential, not only for our personal satisfaction, but also for the survival of our civilization. Science—especially neuroscience—provides a unique window into human nature. Most existing books on the brain are certainly fascinating, but they stop short of explaining the deep implications of neuroscience.

Throughout the book, I elaborate on the meaning of the *physical nature* of the mind and the self, which is based on the observation that the physical disintegration of the brain also destroys the essence of what we are. The piecemeal destruction of the soul by physical agents seems to contradict its presumed unity and supernatural nature. Despite the scientific evidence, some outstanding philosophers and mathematicians still believe that the universe is composed of independent physical, mental, and abstract realms, and most people believe in supernatural entities, such as gods and spirits. Others, including some mathematicians, philosophers, and artists believe in *abstract objects* as if they were independent from the physical world and from the brain. The roots of my determination to find the unity of knowledge were deep, but some questions kept nagging me.

In the midst of my intellectual problems, I realized that faculty members of New York University could attend, with permission of the course director, any lecture at the University. For a couple of years I thus attended some lectures given by professors Ned Block, Paul Boghossian, and Stephen Schiffer in the Department of Philosophy. They in turn frequently had invited speakers, selected from the who's who in philosophy. As a result, I was able to attend lectures and seminars presented by Daniel C. Dennett, John R. Searle,

Paul M. Churchland, Patricia S. Churchland, Susan Haack, Colin McGinn, Robert Nozick, David J. Chalmers, Thomas Nagel, Saul A. Kripke, Susan Blackmore, and many others. I am grateful, therefore, to all the professors that allowed me to come to their lectures, and helped me to navigate through some of the deep waters of philosophy.

Reference

1. Snow CP. The Two Cultures. Cambridge: Cambridge University Press; 1993.

Acknowledgments

So many persons, newspapers, and institutions helped me to write this book that it would be impossible to name them all. My friends and relatives helped me with guidance and advice, but also some adverse circumstances promoted my intellectual development and curiosity. I am indeed grateful for my many outstanding teachers and liberal clinicians that I had in Medical School, including Dr. Eduardo Mancino and Dr. Felipe Cieza-Rodriguez. They did not object to my hospital attendance while I was suspended from the National University of La Plata Medical School because of my political activities against the military dictatorship of Juan Perón in Argentina. During this period, I continued to attend the municipal hospital as if I were a regular medical student, and acquired a solid clinical background. Dr. David Ziziemsky, a young neurologist and psychiatrist whom I helped to administer electroshock treatments to severely depressed patients, took me under his wing and taught me the basic elements of psychiatry and neurology.

After arriving in the United States of America, I had one year of additional training in psychiatry at the St. Louis State Hospital, in St. Louis, MO, and then I became neurology resident in the department of Dr. Morris Bender at Bellevue Hospital in New York. Dr. Bender was a famous neurologist and a keen observer. Presenting a patient to him during grand rounds was a challenging experience. His famous question was "How do you know? This was followed by a screeching "Show me, show me!" which was a challenge because neurological patients during the acute phase of a disease tend to change from one examination to the next. I am grateful that he taught me not to open my mouth until I was sure that the early symptoms on admission were still persistent.

In the course of my neurology training, I became interested in the mechanism underlying the action of drugs on the mind, and I am particularly indebted to the guidance I received while working for 2 years in psychopharmacology with Dr. Menek Goldstein, in the Department of Biochemistry at New York University Medical School (NYU). I continued my training for three further years with Dr. Irwin J. Kopin and Dr. Seymour Ketty at the National Institute of Mental Health (NIMH) in Bethesda, MD, where I was fortunate to meet Dr. Ross Baldessarini, Dr. Solomon H. Snyder—who was still a student—and Dr. Jacques Glowinski, both of whom were working with Dr. Julius Axelrod. I also met Susan D. Iversen and Leslie L. Iversen, a husband and wife Ph.D. team, that later became prominent members of Oxford University. We were all involved in studying the metabolism of catecholamines, such as epinephrine and norepinephrine, so we had many long discussions that contributed to my maturation as a scientist. Dr. Axelrod later won the 1970 Nobel Prize in Physiology and Medicine for his work on catecholamines. After completion of my training at NIMH, I returned to NYU as an Assistant Professor in Pharmacology to set up my own laboratory. During this time, I also managed to attend some philosophy lectures in the NYU School of Humanities, through which I acquired a mixed background that served me well.

The writing of *"Contradictions"* was not an easy task, and I am extremely grateful to all those who helped me. First, I thank my wife Virginia M. Pickel, Ph.D., (Jenny) not only for her emotional support, but also for her active involvement in reviewing and making editorial suggestions about several chapters. Jenny is Professor of Neuroscience at Weill-Cornell University Medical College; she is a prominent neuroscientist who has published over 250 peer-reviewed articles dealing primarily with the structure and immunochemistry of the brain. I also thank my old friend from Argentina, Otto T. Solbrig, Ph.D., a botanist who came to U.C. Berkeley, CA, and then moved to Harvard University, where he became the Bussey Professor of Biology. He made a very helpful critique of an earlier version of my manuscript. I am also indebted to my two daughters, Michèle Musacchio and Andrea Barnhill for their support and encouragement throughout the writing of the book.

I am grateful to Danielle C. Robinson for making all the book illustrations, with the exception of three of them, which are original drawings made by one of my patients. Danielle has a B.A. and a special ability to represent complex concepts with only a few lines; she is now a graduate student at the Neuroscience Graduate Program in Oregon Health and Science University. Her career goals are to combine her artistic talent and scientific knowledge in an effort to obtain a new perspective on normal and abnormal brain functions.

My special gratitude goes to Dr. Angela Lahee, from Springer DE, who was immediately interested in my book proposal, encouraged me to have it published, and provided valuable advice. Ms. Claudia Neumann also helped with many of the details in getting the book printed. I am deeply indebted to Stephen N. Lyle for his expertise and patience in editing my book. I also thank my friend and physician Dr. Philip K. Moskowitz, for keeping me healthy for the last thirty years. I am grateful to Dr. Herbert H. Samuels, Chairman of the Department of Pharmacology of New York University School of Medicine, Langone Medical Center, for providing hospitality and lending support to my efforts, even after my retirement.

New York, NY   José M. Musacchio
April 2012   E-mail: jose.musacchio@nyumc.org

Contents

List of tables

List of figures

1 The universality of religious beliefs

Summary There is a near universality of religious beliefs throughout the history of civilization, a highly puzzling cultural observation that is still true in the twenty-second century. However, reasonable explanations for this universality can be derived by considering the origin of such beliefs. Since religions were an essential cultural element for primitive humans, the practice of imprinting them during early childhood was never abandoned, and religions became an integral component of most civilizations. The practices of religions have changed, but their moral teachings are still based on the Golden Rule, which was already known to Confucius, several centuries before the arrival of the Christian and Moslem prophets. The main religious denominations, however, have lost their unity because they have divided into smaller groups. The varieties of religious beliefs contradict each other, and they also contradict the unity of truth.

1.1 The birth of religions and human culture

The ancestral association of humans with religions is a remarkable cultural phenomenon that cannot be easily overlooked, because religious beliefs are still essential components of all cultures. We can assume that when humans started to wonder about their origins and ultimate fate, they did not have any solid knowledge on which to base explanations for the most basic questions about life and death. Besides, they could not even communicate all their thoughts before developing a complex vocabulary. Thus, they must have lived in a world full of mysteries and unpredictable processes that they could not even name, much less understand, and most groups must initially have relied on their own instincts and on the ways they learnt from their ancestors, who themselves will have learnt from their survival experiences.

J.M. Musacchio, *Contradictions: Neuroscience and Religion,*
Springer Praxis Books, Popular Science, DOI 10.1007/978-3-642-27198-4_1,
© Springer-Verlag Berlin Heidelberg 2012

The complexity of reality was beyond their understanding, so supernatural explanations may initially have provided the only answers to the many questions about their origin and fate, as well as predicting the weather and seasons, which might initially have seemed mysterious. For example, the Sumerians, Ancient Greeks, and Egyptians used supernatural hypotheses to explain the world and invented different gods with the express purpose of explaining specific natural events.

In analogy with children, the thoughts of our early human ancestors probably took place even before they developed an appropriate language, which, according to the most recent theories, took place about 80,000–160,000 years ago in southwestern Africa [1]. Thus, the origin of their world and of their own existence was mysterious, but after their language developed, they probably invented a variety of stories and explanations that later evolved into legends and religions. Stories about the origin of the initial group are likely to have developed long before they could envision any complex form of religion. Even if the stories were imaginary, they were probably told by the elders and had the effect of increasing the coherence of the group, providing a decisive advantage in the ruthless and competitive environment in which they lived.

The widespread belief in supernatural explanations that are fundamental to all religions may largely reflect their naive simplicity, which is easily acceptable to children and primitive humans. Moreover, in the absence of writing, the stories supporting these beliefs were orally transmitted, so they were easily spread and embellished by the elders. The stories underlying religious convictions had the great advantage of providing a measure of hope, security, and protection that is very much needed by all primitive cultures and by children during the early stages of their development and during the difficult times when they have to care for their own families. Children are periodically insecure and need reassurance that they are loved and protected, not only by their parents and the group, but also by some higher beings, which are assumed even more powerful than their own parents. Religions throughout civilization assure believers that they are not alone in this world, and that they will be cared for by some mythological supernatural power. Despite our scientific and philosophical knowledge, it is still difficult to accept that we are alone in the Universe and that our self will vanish with our death.

Most religions are characterized by the belief in a supernatural creator and by the underlying assumption that, through prayers and meditation, humans can somehow communicate with the gods. Even though there are no witnesses to this form of communication, there have been several self-appointed prophets, shamans, and preachers that report trance-like experiences in which they claim to have been contacted directly by a god or by higher spirits. Priests and preachers have also been emphatic in invoking the sacred books written in the name of a god to justify themselves and to justify their authority for dispensing advice. Religions have been used to answer the most elemental questions, often similar to those that are typically asked by children, and to impose a code of behavior, that is also imprinted in them.

Even though the existence of any kind of god has never been proven, religions are based on faith in a god that provides a system of values and beliefs, complemented by a series of rules of behavior that aim to make life in a society organized and predictable. As pointed out by Adler [2], the Golden Rule, the principle that we should behave toward others as we would like others to behave toward ourselves, has been independently discovered by several cultures, and tacitly incorporated into most religions. The universality of the Golden Rule, which was popularized by Confucius (551–479 B.C.E.), long before the Christian Commandments, is a clear indication that moral behavior originated, not in any specific religion, but in a *basic biological need that transcended all cultures and beliefs*. Actually, long before Jesus Christ started to preach, Confucius had already proposed morality in the absence of religious belief. All societies need rules for individuals to be able to live together in a tightly bound community. Unfortunately, the Golden Rule is not always applied to individuals *outside* the clan or the tribe. However, because of the efficiency of modern communications, the size of the group to which our moral rules are applied keeps growing with the advances in civilization. We may hope that the application of the Golden Rule toward different ethnic and political groups will soon become universal.

What seems to be essential for religious dissemination is that religions should be taught in a dogmatic fashion to young children, when they are credulous and can be imprinted for life. Actually, there is a deep analogy between the trust that children and primitive human groups have for their parents and their elders. Older adults who have not been indoctrinated during childhood are not religious or may become religious only under exceptional circumstances, for example, after accidents in which the odds were heavily against their survival. However, the most essential component of religions is the assumption that humans have a soul, a supernatural spiritual component that is the essence of a person, and which will transcend the death of the flesh. Most people understand that their bodies will die and decompose after their death, but believe that their supernatural soul will separate from their body (Fig. 1.1).

Patients hope that their souls will then be received by God in Heaven, where it will meet all their beloved friends and relatives. Some even believe in "the resurrections of the flesh," which as Miguel de Unamuno has remarked, is much more desirable than a purely Platonic resurrection [3].

Religions give a feeling of belonging to a larger group that provides a sense of togetherness, security, and collective meaning, features that are enhanced during religious ceremonies. The universality of religious beliefs has triggered the curiosity of several social scientists and neurobiologists, to the point that some have suggested that humans may have a brain center for a higher

Fig. 1.1 Hallucinations recalled by a patient. Artistic representation of a hallucination reported by a patient after she recovered from a serious disease that produced a near-death experience, in which she felt that her "spirit" was separating from her body

authority or paternal figure, which when occupied provides a sense of confidence and security [4, 5]. Most of us felt secure during our childhood, due to the protection given by our parents. Even later, when we were growing up, we still looked to our parents for guidance. However, when we became too proud to ask for their help, or they could no longer provide this support, we needed to become self-sufficient or get somebody else that would take their place. Perhaps the memory of them or the example they set gives us a sense of confidence and security which is later transferred to a supernatural being that symbolically takes their place. We do not like to be alone and unprotected, but our mere desires are obviously not sufficient to create any god.

1.2 The evolution of religious beliefs

It would not be possible to cover all our knowledge of primitive cultures and their supernatural postulates in an adequate way in this brief overview of religious beliefs. Moreover, different cultures have evolved in different fashions, because the sequence of their religious development is not as rigid as the stages of biological development. However, by identifying the key developmental stages that have culminated in today's religions, it may be possible to summarize vast amounts of information that could help us to understand the deeper roots of religious beliefs. Religious practices may progress from the worship of natural forces and elements, through animism, ancestor worship, totemism, and shamanism in various orders, and finally culminate in different forms of theism and polytheism. All these practices support supernatural beliefs that involve magic and multiple deities, as will be discussed below.

Primitive cultures, like infant humans, tend to believe that natural processes such as rain and storms are *animated* or are actually the work of somebody, as several researchers have pointed out [6]. It is clear that early humans were not scientists, but their logic and primitive technology taught them that things occur due to a cause, and they were able to anticipate the dangers of signals indicating the presence of a predator or natural enemies. However, natural processes without obvious causal agents were often attributed to supernatural powers and to minor deities that had specialized functions.

The cult of the ancestors, veneration of the dead, and burial rituals were practised by almost all ancient cultures, not only because they were thought necessary to heal painful experiences, but also to preserve the souls of the dead. These rituals involved praying, and some cultures also had altars at home, where the ancestors were offered food and prayers, a practice that is still common in several oriental countries today.

Totemism is another a primitive practice involving a mystical relationship with totems, which are animals, plants, or objects used as the emblem of the group and considered sacred. Totemism includes a collection of different beliefs, sometimes involving the cult of ancestors and the dead, belief in spirits, and certain taboos. Breach of these taboos, or an offense or damage to the totem, may bring sickness or death to individuals or to the whole group. Totemism is rich in rites and ceremonies involving dances and imitation of animals, humans, or natural processes. It was claimed by the French sociologist Emile Durkheim (1858–1932) that totemism was a fundamental characteristic of all religions, but this is no longer accepted today. In contrast with totemism, *shamanism* is an animistic system of beliefs in which mediation between the visible and supernatural worlds is effected by shamans. These are individuals who claim to receive their powers directly from the gods and spirits, and acquire their status through personal communication with the supernatural. Such communication takes place through dreams and visions, some of which are believed to be produced by visits from the souls of dead or living persons, or by other spirits [7].

Some anthropologists take the view that religions emerged largely because of the limited "successes" of magic in dealing with natural disasters, and as a confession of human impotence in the face of death or natural forces. These limitations created the need to believe in and recruit the help of higher powers through religion, which was and still is a fundamental component of human life. However, much of the information on supernatural beliefs and worship by illiterate tribes and cultures has been lost forever. To fill the gaps, it has been assumed that their beliefs were similar to those documented in the nineteenth and early twentieth centuries by numerous anthropologists who studied primitive cultures that remained isolated in remote parts of the world. Despite some contention, it is generally agreed that there has been a continuous progression from supernatural and superstitious beliefs to the more complex and sophisticated religions.

Religions developed in parallel with civilization, long before recorded history. Like many of today's primitive cultures, our early ancestors may have used prayer, dance, and complex ceremonial ritual to communicate with the gods or with the spirits thought to control their destiny. Some primitive Melanesians believe in a mystic impersonal power, which is called *manna* and has equivalents in other primitive cultures. In their efforts to control the environmental factors required to successfully cultivate their gardens, or to fish and hunt, primitive societies exploited their practical knowledge, but they often supplemented this knowledge with magical practices to counter the unexpected and uncontrollable natural forces that could transform a well-executed operation into a disaster. Primitive cultures knew very well which processes could be controlled by applying the appropriate techniques, and which were out of their control and had to be prevented if possible using magical ceremonies. Similarly, our early ancestors may have resorted to begging the help of the superior powers attributed to the gods only after recognizing the inadequacy of their magic and mystical solutions for daily problems.

Offerings are a common component of religious rituals in which something valuable is given to the gods or spirits as a conciliatory gesture or homage, the most extreme being the ritual sacrifice of an animal or a person. It is interesting that myths, rituals, and sacrifices have developed in all primitive cultures, apparently in an independent fashion. Even the most primitive people seem to recognize that, to get something from somebody, you have to offer something in exchange. The anthropomorphic character of the gods is implicit in the assumption that they will appreciate the offerings, even if the offerings are useless to the gods, and only represent a sacrifice or hardship for the givers.

The birth of myth and religious belief seems to be directly related to our constant search for meaning and to our human fear of death. According to anthropologists, early humans would have assumed that human experience did not end after the death of the body, so they conceived the existence of a creator who gave them a spirit or soul that would survive death. They also left indications that there were good and bad spirits who could affect their life. Similar beliefs may have developed independently in many isolated populations throughout the world, suggesting that the idea of extinction after death seemed unacceptable to all early humans in different civilizations. It also indicates the birth of symbolic thought and the use of symbols and symbolic behavior, such as rituals, to transmit meaning.

Ernst Cassirer rightly considered that the most characteristic human attribute is the abstract capacity to invent and use symbols [8]. The earliest record of symbol writing has been found in ochre engravings of unknown meaning in the Blombos Cave in South Africa, dated to about 100,000 years ago [9]. Ice age hunter-gatherers who lived 30,000–50,000 years ago painted, not only realistic animals, but also some half-animal and half-human figures in several European caves. The significance of their drawings, or their carvings of little ivory statuettes, remains unclear, but they are certainly examples of a burgeoning artistic ability. The religious meaning of many sculptures in Mesopotamia and Egypt only became evident to us much later, with the decoding of writing, which was invented about 5,000 years ago.

The oldest known temple was built in Göbekli Tepe, southern Turkey, about 11,600 years ago. In fact, several temples have been found in a large area that runs from southern Turkey to the southeastern side of the Fertile Crescent, near the mouths of the Tigris and Euphrates rivers in the Persian Gulf. The temples contain several pillars or standing sculptures representing humans. These are made from a single piece of stone 6 m tall, and are surrounded by circular walls, with a long aisle that provides access. The pillars also bear carvings of several local animals such as boars, cranes, foxes, scorpions, and snakes. The area was fertile and provided an ideal setting for prehistoric farming, as attested by the discovery of domesticated wheat and other grains. There is little doubt that this area became the cradle of civilization, even if it received influences from several different directions.

The complex religious beliefs of different cultures reflect not only their mythology, but also the stage at which they stabilize or cease to evolve. Local conditions imposed by different natural forces, food abundance, or periodic floods also contributed to determining the forces that were feared and worshipped by the different cultures. Early humans continually sought to control the forces of nature for practical purposes, and to avoid natural disasters, epidemics, and drought. The heavens are generally considered to be animated, since they *seem* to be the source of rain, wind, lightning, and thunder. Thus, many primitive cultures believe in heavenly gods who control to varying degrees the events that affect their lives. Interestingly, some of these beliefs, such as the exact location of gods, angels, and other minor deities in heaven, are still essential components of today's major religions.

Religions are now found throughout the world, and are almost universally hailed as the highest expression of the human spirit. The early supernatural beliefs, however, were varied and heterogeneous, because they were polytheistic, with different gods and spirits who fulfilled specific functions for which they were worshipped. Animism still predominates in some illiterate cultures. It consists in the belief that, in addition to humans, animals and other objects also incorporate a spirit that somehow provides their essence. They also believe that during sleep, as in death, human souls can separate from the body (Fig. 1.1) and visit other persons, in whom they produce dreams.

Early religions started out polytheistic, as though every important natural element had to be worshipped independently to help struggling early civilizations. Other conceptions of the supernatural include pantheism, the belief that every object in the Universe is sacred, and theism, the belief that one god created and rules the Universe. The latter should be contrasted with deism, which is the belief in a god based on reason rather than revelation, and in particular a god who created the Universe, but does not interfere with humans and does not control the Universe.

1.3 Western religions

Whereas there are many different types of religion in the world, there is a tendency for underdeveloped cultures to believe in polytheistic religions. For example, early civilizations like the Greeks had many deities that were classified in different categories and had specific functions. Zeus was their main god, but there were many others, such as Aphrodite, who symbolized love and beauty, Prometheus, the creator of humankind, Athena, the goddess of wisdom, and Dionysus, the god of wine. The Romans, Egyptians, and Phoenicians also had a large collection of major and minor gods.

Today, the most widely followed religions are Christianity, Buddhism, Islam, and Hinduism. These are divided into several different branches, spread through different regions and countries. Judaism is actually a small religious/ ethnic group that is only politically important in the Middle East and in the United States.

Christians believe in Jesus Christ as their main deity and in the writings of the New Testament. Jesus is thought to be the Son of God, who was sent to Earth to save humanity. Christianity spread rapidly in different countries and regions, so it is no surprise that Christians eventually divided up into several groups. And since they spread in three different directions, they split into three main subgroups: Roman Catholic, the Eastern Orthodox Church, and Protestants. These later subdivided further into various churches.

Islam is also a monotheistic religion that believes in Allah (or God). Like the Christian God, Allah is thought to be the creator of the Universe and the source of all good and all evil. Muhammad (570–622 A.D.) was the most important prophet of Islam and codified the strict rules that Muslims are required to follow today. Like Christians, Muslims believe that, after death, they will enter either paradise or hell. They must do a pilgrimage at least once in their lifetime to worship at a shrine in Mecca. Despite the similarity of their expectations after death, there are important differences in the beliefs imposed by these religions.

In some parts of the world, Buddhism is seen, not as a religion, but as a philosophical system. Based on the teaching of Siddhartha Gautama or Buddha, it encourages followers to free themselves from the tyranny of materialism.

Other religions have developed in Africa and the islands of the Pacific Ocean, and among the indigenous populations of the American continent. There are mixtures of aboriginal, Caribbean, and Christian religions that prosper today, but it is not our purpose to detail all the variants.

The Sumerian, Egyptian, Roman, and Greek religions venerated several gods, each of which provided protection in specialized areas. In contrast, Christianity, Judaism, and Islam are monotheistic religions, which worship a single personal god who is the creator and ruler of the Universe and actively intervenes in the personal affairs of humans and communicates with them. As indicated previously, deism is the belief in a god that created the Universe, but does not intervene in human affairs, while an atheist is someone who does not believe in any personal god, and an agnostic is someone who is not convinced one way or the other about whether gods exist.

In America, there is a diversity of local traditions and religious beliefs that are thought to be intrinsically respectable and that are never openly criticized by political leaders, even though they contradict each other, and even though, in the best of cases, only one of them could be completely true. In the average American city and in the countryside, there are many competing churches showing that religious freedom and practice, within socially acceptable limits, often forms the backbone of small communities. Religious participation is almost obligatory, because it procures a respectable status in the community and a social outlet for interaction, not only at a personal level but also in business and politics.

This situation is not always constructive, because in some parts of the country, political and religious leaders control the decisions of school community boards and oppose the teaching of biological evolution, for example, which is unacceptable to most religions. Community boards also oppose the innovations brought about by competitive newcomers and immigrants.

Thus, while the universality of religious beliefs can be explained when we look at their primitive origins, their fragmentation suggests that they are all affected by local legends and political groups. Religions have survival value when they serve to bring together and promote coherence in groups of people, but they have also been a major source of conflicts, some of which are still raging around the world, with little hope of appeasement.

1.4 The detachment of the spirit

Tabloids and popular magazines often publish descriptions of near-death experiences that require the subject to be in an extremely abnormal organic state. People get sick and approach death only when there is something seriously wrong with their bodies. Under normal conditions, the composition of the body fluids is kept within narrow limits, what Claude Bernard called the *internal milieu* [10]. When the composition of the body fluids is altered by disease, the functions of most vital organs such as the heart, kidney, liver, and brain are deeply disrupted and this can produce a whole range of organic mental symptoms including delusions, hallucinations, and delirium.

Delirium is the most frequently observed mental disorder in clinical practice [11]. Manifestations include reduced awareness and attention, memory impairment, disorientation in time and space, rambling speech, and incoherence. Delirium may also involve many associated features such as anxiety, fear, depression, and perceptual disturbances, including illusions or hallucinations. As indicated below, delirium can be caused by almost any factor that alters brain function. About *10%* of medical-surgical patients become delirious at some point during their hospitalization, but the incidence increases to *40%* in geriatric patients. Delirium due to a general medical condition is characterized by disturbances in consciousness that cannot be explained by a brain lesion, and it is strongly associated with increased risk of death. Delirium characterized by vivid dreams and hallucinations is frequent in children and other individuals with high fever and dehydration.

In short, a variety of conditions that alter the composition of the blood beyond certain limits can produce an abnormal state of consciousness and eventually cause death. Delusions of detachment of the spirit from the body (Fig. 1.2) are frequent during periods of prolonged dehydration, intoxication by drugs, and severe organic disease, as experienced during agony.

The evidence indicates that near-death experiences are *organic mental syndromes* that fall under the category of delirium. Normal healthy individuals do not have near-death experiences, but the number of patients that recover from near-death experiences has increased during recent years because the high technology of modern medicine can save the life of a patient that previously would certainly have died. The typical cause is cardiac arrest followed by coma and cardiopulmonary resuscitation. However, near-death experiences are reported only by a small proportion of patients that undergo cardiopulmonary resuscitation. Delusions of detachment of the spirit from the body, so typical of

Fig. 1.2 Hallucination of observing her own dying body. A different patient reported after her recovery, a near-death experience with hallucinations, in which she felt that her soul was observing her own dying body and members of her family who were taking care of her

near-death experiences, are also produced by hallucinogenic drugs, including mescaline and LSD.

The diversity of the descriptions provided by patients that recovered from these extreme pathological conditions indicates that their visions were conditioned by their preconceptions, culture, and religion. Their stories are inconsistently described by different persons and in reports from ancient times, as in many biblical revelations and visions. The vividness of the descriptions and the depth of the near-death experiences obviously do not provide any indication of their reliability. Unfortunately, some of the reports on near-death experience have been performed just by collecting information provided by patients that anonymously answered a mailed questionnaire, without describing the physical condition that caused the experience [12]. However, the proposal that the near-death experiences could be labeled as "religious or spiritual" ignores the fact that such experiences usually take place in semi-comatose or delirious patients with serious organic abnormalities. The most ardent believers are often individuals hoping for any sign that could be interpreted as an indication that they have a soul that will survive death.

References

1. Atkinson QD. Phonemic diversity supports a serial founder effect model of language expansion from Africa. Science 2011;332(6027):346–349.
2. Adler JM. Truth in Religion. The plurality of religions and the unity of truth. 1st. edn. New York: Macmillan Publishing Co.; 1990.
3. de Unamuno M. The Tragic Sense of Lie in Men and Nations. First Princeton Paperback edn. Princeton: Princeton University Press; 1977.
4. Snyder SH. Seeking god in the brain – efforts to localize higher brain functions. N Engl J Med 2008;358(1):6–7.
5. Trimble MR. The Soul in the Brain: The cerebral basis of language, art and belief. 1st edn. Baltimore: The Johns Hopkins University Press; 2007.
6. Culotta E. Origins. On the origin of religion. Science 2009;326(5954):784–787.
7. de Waal Malefijt A. Religion and Culture. An introduction to anthropology of religion. 1st. edn. London: Collier-Macmillan Limited; 1968.
8. Cassirer E. An Essay on Man. An introduction to a philosophy of human culture. 1st. edn. New Haven and London: Yale University Press; 1944.
9. Balter M. Human evolution. Early start for human art? Ochre may revise timeline. Science 2009;323(5914):569.
10. Bernard C. An Introduction to the Study of Experimental Medicine. New York: Dover Publications, Inc.; 1957.
11. Horvath TB, Siever JL, Mohs RC, Davis K. Organic mental syndromes and disorders. In: Kaplan HI, Sadock BJ, editors. Comprehensive Textbook of Psychiatry/V. 5th edn. Baltimore: Williams & Wilkins; 1989.
12. Greyson B. Dissociation in people who have near-death experiences: out of their bodies or out of their minds? Lancet 2000;355(9202):460–463.

2 The contradictions and consequences of religious beliefs

Summary Humans have assumed for thousands of years that they have a spirit or soul that will survive the body after death. These beliefs gave rise to countless religions that developed in different directions and led eventually to Christianity, Islam, and Judaism, and many minor religions. Hinduism and Buddhism, however, could be better described as a mixture of philosophy and religion. All religions are based on faith, so not surprisingly they evolve in different directions and originate countless branches and subdivisions. It should be noted that these contradict not only each other, but also scientific knowledge. Faith is a primitive form of irrational knowledge in which the truth or falsity of the beliefs cannot be tested. But faith is an essential component of all religions.

Drastic religious changes took place during the Reformation in the sixteenth century, and some of these had major political and religious consequences, sparking off many political conflicts and still fuelling wars even today, mainly in the Middle East and in western Asian countries. The freedom given to all established religions and new religious movements has unintentionally created further conflicts, many of which have ended in tragedy. Extreme faith can produce shared psychotic disorders that may have lethal consequences.

2.1 Cultural and religious contradictions

The first encounters with contradictions probably originated at the dawn of civilization, when humans started to face the contrast between their raw instincts and the incipient human aspirations of having a shelter and food for their family. Our hominid ancestors (Chap. 3) were probably striving to separate themselves from the ruthless animal world in which they were born and to avoid the constant challenges of dealing with predators and competing hominid groups. The evolution of the hominid brain progressed slowly, but the development of the prefrontal cortex and the areas of association that increased the interconnectivity of different brain regions and functions resulted in a dramatic increase in intelligence and in the probabilities of survival. These changes were mainly due to development of areas related to creative intelligence and language.

With the progressive development of higher intelligence, abstract thinking, and language, primitive humans became better prepared to face life, and as agriculture gradually came on the scene, there was some free time between harvests to think and to wonder about the meaning of life and death. Early humans assumed at some point that there must be something else after death [1–3], so they conceived the existence of a supernatural creator and assumed that they had their own supernatural soul that would survive death. In analogy with their own world, they thought that there were good and bad spirits, angels, and devils, who could affect their lives. Similar beliefs apparently developed independently in many isolated populations.

Today, the major religions are Christianity, Islam, and Judaism, if we consider Hinduism and Buddhism to be better described as philosophies of life rather than religions. The Christians believe in Jesus Christ as their main deity, while followers of Islam believe in Allah. Like the Christian God, Allah is believed to be the creator of the Universe and the source of all good and evil. Muhammad (570–622 A.D.) was the most important prophet of Islam; he codified the strict rules that Muslims must follow, which include praying in Arabic five times a day. Besides, during Ramadan, the ninth month of the Moslem calendar, Muslims must abstain from food, drink, sex, and smoking, from sunrise to sunset. They must go on a pilgrimage at least once in their lifetime to worship at a shrine in Mecca. At death, they believe, like Christians, that they will enter either paradise or hell.

But despite their similar expectations after death, these religions involve important differences that preclude their unification into one single religion. Thus, since the main religions were born, there have been inherent contradictions between believers. The existence of contradictory beliefs between different cultures is one of the reasons why the world is chaotic, generating arguments about politics, territories, and religion which can easily result in armed conflict and religious persecution. Unfortunately, the United States government sometimes feels compelled to intervene as a peacemaker to protect American investments.

The American political, economic, and racial confrontations of half a century ago were solved with minimal violence. Today, we have in theory the freedom to choose beliefs, express opinions, and decide how to live our lives. While these are considered among the most important assets for American people, only a small fraction of the population really has the means and the privilege to choose their career and attend the schools of their choice.

Even though our freedom is nominally guaranteed by the constitution, the choices available to most Americans, and especially of those that have just immigrated, is unfortunately limited by a constellation of cultural, social, and economic factors. They include the cost of higher education, the lack of popular role models outside the sports or musical arenas, and the provincial mentality that predominates in many small communities.

One of the most severe blows to American confidence was produced by the assassination of President John F. Kennedy. The shock of losing our hopes for rationality and innovation was compounded by the passive and uncritical acceptance of family tradition, religion, and social conventions. Still today, freedom of choice cannot truly be exercised by most people, for they become prisoners of their own traditions and environment. In addition to their poor education and the need to work long hours to pay for food, high-interest mortgages, and the traps of easily obtainable credit cards, the average American worker is constantly *distracted* and *diverted* from preparing for a better future. There are no Roman circuses with gladiators, but there is football, basketball, and baseball, not to mention video games and movies that provide immediate entertainment to *divert* a large sector of the population from more fundamental issues. Americans seem to live in a modern adaptation of Huxley's *Brave New World*, with a blind and irrational confidence in a future inspired by television shows, soap operas, or sporting matches.

We have a diversity of local traditions and religious beliefs that are all thought to be intrinsically respectable, and cannot be openly criticized by political leaders. In the average American cities and in the countryside, there are many competing churches showing that religious freedom and practice are the backbone of our society. Religious participation provides a respectable status in the community and a social outlet to interact, not only at a personal level but also in business and politics. These developments are not always positive, because political and religious leaders are beginning to control the decisions of school community boards, and to oppose the teaching of biological evolution, for example, as well as the innovations brought about by competitive newcomers and immigrants. To make the situation worse, the voice of intellectuals cannot be clearly heard over the background noise created by primetime television commercials and political propaganda [4].

Our cultural tradition also seems to have been short-lived, because many nonfiction authors whose books were read by the previous generation are hardly read today. Interestingly, books for or against religion are selling well today, but the great majority of nonfiction books are written about popular entertainers, sports heroes? and political personalities. There are also many "How-To" books, on cooking and other crafts, as well as travel, dieting, adventure, and other entertainments. But the number of scholarly books produced by private publishers is actually at an all time low. Even university presses often agree to publish serious non-fiction only if the book could be used for teaching.

This is because non-fiction books simply do not sell nearly as well as popular fiction or memoirs about the horizontal lives of famous actors. Perhaps the new electronic forms of communication and the WWW will facilitate the reading of thought-provoking books and replace the role of the classic *public intellectual* that characterized Europe a few decades ago.

The cultural crisis is not new, but it has reached unprecedented proportions. As Adler indicated, it was Averroës (1126–1198) who proposed to accept independently the "truths of faith" and the "truths of reason". However, he assigned a higher status to the truths of reason, because the truths of faith belong to the sphere of imagination [3]. Besides, Averroës thought that religious writings could be interpreted allegorically, giving him room to avoid contradictions. Averroës was probably the first to suggest life in such a schizoid world. Dualism reaffirms the existence of different realms and the use of faith as a legitimate method to validate beliefs.

Descartes (1596–1650) contributed significantly to putting dualism on apparently solid grounds [5], and other philosophers and some scientists have tried to justify the simultaneous acceptance of two independent realms. This was probably a way to avoid conflict between religion and reason, which could have been fatal during the Middle Ages or the Inquisition. Even today, some philosophers and scientists keep their religious beliefs in a compartment well separated from their objective knowledge. The majority of people get along with social conventions by dividing reality into two independent worlds, the *natural* and the *supernatural* [3]. The objective world is understood through science, which relies on reason, logic, and empirical knowledge, while the supernatural world, ruled by *faith*, provides religion, astrology, and other superstitious beliefs. The duality of body and spirit, which is implicit in today's popular culture, is still fundamental to the organization of the intellectual and spiritual life of most people.

2.2 Early spiritual beliefs

Homo sapiens seems to be anything but *sapiens* in relation to faith and religious beliefs. Our early ancestors may have started to have religious thoughts when the Neanderthals began to bury their dead in special positions. Burying a parent or a child is certainly a painful experience. It feels like a deep injustice, and makes us want to take good care of the dead, compelling us to ponder the meaning of life and death. The existence of these universal feelings suggests that religions would have been developed in the early stages of civilization by the most primitive and ignorant cultures, who still believed in magic, spirits, and witchcraft. As discussed below, people thought that being alive and moving, that is, being *animated*, meant having a soul.

All religions support the idea that humans have a soul or a spirit. Animals were also thought to have a soul, *anima*, or *vital spirits*, because they can move at their own will. Actually, the word *animal* derives from *anima* or soul in Latin. Observing the movements and autonomy of animals fascinates everybody, especially children. This is probably why Walt Disney's *animated* cartoons became so popular, and today animation is used in television, not only in advertisements for children, but also for adults. The existence of animal spirits, however, was disproven in the 1700s, when the Italian physician Luigi Galvani (1737–1798) was able to produce movements in the legs of recently sacrificed frogs by applying electrical discharges to the frogs' sciatic nerve.

The use of electricity to study animals has had far-reaching scientific and philosophical consequences, besides disproving the theory of "vital spirits" to explain animal movements. Vital spirits were replaced by what Galvani called "animal electricity". Galvani's experiments were published more than two centuries ago, in 1791. They provided the basis for the studies of Alessandro Volta (1745–1827), who invented the battery, and for the German physician Emil du Bois-Reymond (1818–1896), who continued to study animal electricity in electric fishes. The multiple devices designed by these investigators to register animal electricity led many years later to the invention of the EKG or ECG (electrocardiogram) and EEG (electroencephalogram), and these in turn have provided the starting point for much more sophisticated imaging techniques, such as positron emission tomographic scanning (PETS) and functional magnetic resonance imaging (fMRI). Besides being powerful diagnostic tools, these can visualize some of the brain activity involved in mental processes and in the maintenance of consciousness.

2.3 The contradictions between science and religion

Most cultures have been through similar, but not identical developmental stages that started with primitive taboos, animism, and polytheism. The evolution toward monotheism seems a major simplifying step. While replacing the complexity of polytheism in some cultures, it required a greater capacity for abstract thought. However, some religions that are considered monotheistic, such as Christianity, actually refer to a multitude of deities, including various saints and virgins, who are said to be specialized in solving specific problems or in helping to cure some diseases. In many churches, there are separate altars that allow people to pray directly to different saints. But, if everything else fails, it is still possible to prey directly to any member of the Catholic Holy Trinity, which includes the Father, the Son, and the Holy Ghost. But the idea of three persons existing in One Divine Being is a Catholic dogma that is not acceptable to Protestants. The Catholic Trinity has been declared a mystery, and it has been complicated by more than 2,000 years of theological arguments and irrational explanations.

Today, almost 90% of Americans are religious and believe in two independent realms, living thus in a schizoid universe with a natural and a supernatural component. Moreover, some contemporary philosophers have even tried to demonstrate that physicalism is wrong, proposing that there is a fundamental difference between the subjective character of qualitative experiences and the physicality of the brain [6, 7]. It is a pervasive illusion to believe that qualitative experiences and the mind are nonphysical phenomena. The overview provided in this book will show that qualitative experiences are ineffable, not because they are mysterious or supernatural, but because they consist in *language-independent* neural processes that could be named, but not re-created in the brains of others through explanations (Chap. 4). Actually, languages can only be developed and translated by anchoring words or gestures into qualitative experiences, or by making the appropriate sounds and gestures that are understandable because humans and higher animals have similar phylogenetic backgrounds and common experiences [8] (Chap. 6).

There are no satisfactory explanations for how the natural and the supernatural realms could interact with each other. Do we have a dual nature, material and spiritual as most religions teach, or are the duality of body and spirit only an ancestral hope supported by religious beliefs, family tradition, and wishful thinking? The most common answer is of course that we have a body and a spirit. At some point in our lives, we have all been deeply preoccupied about our nature and ultimate fate. We realized, probably during our early teens, that we are not just acting in a school play, but we are living our real and only life. When we were young, we envisioned growing up, preparing to face life, and realizing all our dreams, while knowing that some day in the very, very distant future we will be facing death. We may also have sometimes hoped that the preachers were right, and that we would enter the supernatural world, where we could live happily through all eternity in the company of our loved ones.

But this would be possible only if we had a *dual nature*, a physical body and a spiritual soul that would survive death. Actually, it really does seem as though we have a spiritual or intellectual component that is not physical. I will show later that this feeling is produced by the impossibility of perceiving our brain and its functions, which are then attributed to a supernatural soul. We have no doubts that our physical body will eventually die and decay, so it seems encouraging that the major religions and our subjective feelings tell us that we may also possess a non-physical spiritual component. The problem with this view is that faith and beliefs are not rigorous ways to tell what is true or false. Faith *is only a primitive way of validating beliefs*, because it gives rise to innumerable contradictions, as indicated by the religious incompatibilities that have always plagued the world. Moreover, faith is not compatible with science and reason, both of which we have learned to trust. Religion and faith require belief in an interaction between the *natural world* and a hypothetical *supernatural realm*, but nobody has explained how the two realities could interact without violating the most fundamental principles of physics and reason.

Belief in the existence of a spiritual nonphysical component implies that there is a natural reality and a supernatural reality, and that we are part of both worlds, even though there is no evidence for the latter. We know that we have a physical body and a thinking mind as well as a strong moral sense and deep emotions that seem somehow spiritual. Nevertheless, there are too many religions and contradictory beliefs for all of them to be true, as each individual religion claims. When we grow up, no answer comes, and regardless of our education or occupation, we realize that we still do not know what comes after death, if anything. Thus, we carry on with dignity, while looking for reassurance. The majority of us initially accepted the most popular belief: if we are good in this world, our souls will go to Heaven after we die, but if we are not, we will go to Hell, where we will suffer forever. However, this statement is just a childish dream, and some of us have serious doubts about the existence of a God with good intentions who would create a perverse Devil. It seems that the God of the Bible was also schizoid.

The most ironic aspect of religions is that they were invented at a time when primitive humans were living in deep ignorance, when the Earth was thought to be flat, and the Sun to revolve around the Earth. Since religions cannot discriminate between false beliefs and true knowledge, they simply qualify as superstitious beliefs, the irrational creations of well meaning, but deeply ignorant people, some of whom were in addition epileptic or paranoid enough to have auditory hallucinations, such as the Prophet Muhammad.

Mary Baker G. Eddy (1821–1910), the founder of the *Church of Christ, Scientist* in Boston, was also mentally unstable and reported that she was able to have conversations with God. She founded the Christian Science Monitor, a newspaper, and a magazine. Her teachings advocated refusing medical treatment in favor of prayers, because she believed in the spiritual healing of diseases, including all the bacterial diseases of childhood. It was said by her followers that she was able to heal people instantaneously, and this could well have been true if she was dealing only with rampant psychosomatic complaints. Initially, she had thousands of followers, but the number of believers has shrunk to less than 10% in recent years. Eventually she was found guilty of contributing to the deaths of several children, who got prayers instead of regular medical treatment. Some of her followers still believe in spiritual healing, and even recently this has resulted in the unnecessary deaths of several children, whose parents are being brought to trial under the accusation of manslaughter.

Since all religions claim to be true, none of their claims can be taken seriously, simply because religions contradict each other. A true religion is an oxymoron. In addition, there are major inconsistencies between the omniscience and the omnipotence of the Christian god and the contradictory realities in which we live. God seems to ignore all the problems that affect humans, including the chronic infantile mortality and the hunger of children in Africa, especially Somalia, in the area of the African Horn, where in August

2011, children less than 5 years of age were dying at the rate of 10,000 per month. It seems reasonable for humans to expect an omnipotent god to solve at least some of the serious problems that chronically affect many children.

In fact, such problems should never have developed. God's neglect is universal, because there are many catastrophic situations around the world. For example, God should have stopped all the calamities that have plagued the world in recent years, such as the devastation of New Orleans by the hurricane Katrina in August 2005, which killed more than 1,800 people and destroyed a considerable fraction of the poorer neighborhoods. Not to mention earthquakes, which regularly produce a much higher toll: an earthquake of magnitude 7 in Haiti killed more than 85,000 people in January 2010, and the Great East Japan Earthquake of March 2011 produced a tsunami that resulted in more than 13,500 people being killed, while 14,500 were still missing a month later. The same tsunami also destroyed three nuclear reactors and resulted in the spread of highly radioactive particles that have been incorporated into the food chain and have made a large area uninhabitable for many years to come. While religious leaders might argue that these disasters are necessary to test the strength of human character, they are much more likely simply to indicate natural forces beyond our control, with or without the help of Divine intervention.

The multiplicity of religions is illogical, unnecessary, and contradictory, because we only need one true religion and one effective god, but neither can be identified with any degree of certainty. Actually, although this critique may sound childish and naive, it applies to the gods of all religions, all of whom have been neglecting the needy around the world. If Jesus Christ was actually a god, who descended from Heaven to be with the mortals, it seems strange that he has not been able to come back to talk to and advise humans for the last two millennia. God should be aware that many serious problems have occurred since his last visit; we have had many wars and plagues, and more recently, we have had two world wars that were much more devastating than anyone could have imagined two millennia ago. Besides, God has also been ignoring the crimes of tyrants and despots from Argentina and Afghanistan to Tunisia. Are these tests of our character and faith, as proclaimed by many religious leaders, or are they an indication that we are on our own and would do better to protect ourselves.

2.4 The fragmentation of culture

The problems of human nature are complex and intertwined, so *Contradictions* is not only about beliefs, but also about the nature and reliability of our culture. Inspired by C. P. Snow's "The Two Cultures" [9], I realized that the arbitrary division of knowledge into disciplines allows us to create tight compartments to avoid contradictory beliefs and confrontations. For example, we could justify almost anything by taking refuge in a particular religion, philosophical doctrine, or narrow academic discipline. The idea is to convince people and ourselves that other disciplines have no relevance to the subject under discussion, or that they have an "independent nature". We easily forget that nothing is in fact independent in this way, because universal laws encompass the whole Universe and supersede our provincial cultures. Besides, matter and energy are physically equivalent as described by Einstein. Opinions are subjective, but truth is universal (or at least it should be), and it has a coherence and beauty that leaves no room for contradictions. Today, the coherence of knowledge and truth are both essential, not only for our intellectual well-being, but also for the survival of our civilization.

Our cultural situation contrasts sharply with the Western philosophical tradition that started with the cosmological preoccupations of the pre-Socratic philosophers, who speculated on the origin of celestial bodies, life, and the structure of matter, the existence of movement or the nature of numbers. The early Greek philosophers were not only theoreticians, but also tried to solve practical problems, such as measuring fields and calculating the distance of faraway objects. With Socrates, Plato, and Aristotle, philosophy expanded into an all-encompassing body of inquiry that included the origin and the possibility of knowledge, ethics, geometry, logic, biology, metaphysics, etc.

Specialization and the ensuing fragmentation of knowledge into disciplines and sub-disciplines is one of the generally accepted, but potentially divisive developments that today threaten the unity of our culture. The fragmentation of knowledge is institutionalized by academic departments at many universities, which sometimes become responsible for the fragmentation of teaching. These departments are organized and maintained for several reasons. Initially, their names were derived from a new specialty with a fancy name, which suggested that their members were working in the forefront of knowledge. However, the dual obligation of research and teaching created the need for a territorial demarcation. Moreover, they have personnel that have to be supervised and sometimes paid for by obtaining competitive research grants. Thus, academic departments are *territories with a budget*, which become symbols of the power of each departmental chairman, who tends to perpetuate the status quo by resisting any attempt to reorganize or to change the curriculum. This is why academic departments are generally reorganized only after the replacement of a chairperson.

2.5 Religious atrocities and shared psychotic disorders

Different churches and new sects start as small congregations that branch out from larger denominations, such as Judaism, Christianity, or Islam. Many developing sects incorporate elements from other religions. For example, Christianity began as a synthesis of Judaism and Greek religion which also incorporated elements from Celtic religions and Roman paganism [10]. All major, time-tested religions have proven their social acceptability and have generated major independent denominations. The explosion in the number of denominations can probably be traced to the Reformation. Ulrich Zwingli (1484–1531), who initiated the Reformation in Switzerland, preached that the Bible is the absolute authority. The German theologian and leader of the Protestant Reformation Martin Luther (1483–1546) also accepted the Bible as the sole source of revelation. He believed that salvation would be granted on the basis of faith alone, and supported a universal priesthood of all believers.

Unitarianism was a movement that emerged from the Reformation and favored a view of the scriptures as interpreted by reason. They did not believe in the Trinity or in the divinity of Christ. One of the first promoters of Unitarianism was the Spanish physician and theologian Michael Servetus (Miguel Serveto), who discovered the circulation of blood through the lungs. He argued that there was nothing in the New Testament that contradicted the monotheism of the Jewish Scriptures [11]. For his denial of the Trinity, Servetus was burnt alive at the stake in 1553 by order of John Calvin. His execution provoked a justified reaction against punishing heresy with death, and this probably had a role in the spread of Unitarianism to England during the seventeenth century. The seditious movements breaking off from the central authority of the Roman Catholic Church, and the license for free interpretations of the Bible which could be validated just by personal faith alone, contributed to the generation of a wide variety of religious denominations, sects, and cults. The Bible thus became a useful kit for developing religions to suit individual tastes and the fashion of the times.

The Protestant legacy and the religious freedom guaranteed by the American Constitution, assure the endless proliferation of sects, churches, and denominations. This proliferation is further fueled by competition between the different denominations, which have to recruit followers and donors to survive. Churches without parishioners do not prosper. Kosmin and Lachman [12] list an impressive number of denominations in their text and tables, but after several failed attempts to count them all, let us just say that the U.S. Department of Defense lists 260 denominations under the general heading of Protestant chaplaincy alone.

2.6 The "new" religious movements and their apocalyptic predictions

Dissident sects or churches are usually recognized as legitimate, even though some are weird, by almost any standard. They are tolerated in the name of religious freedom, providing that they are not socially obnoxious. Sometimes these sects evolve into large denominations, despite certain odd features. An example are the Mormons, who were polygamous until 1890, or the

Christian Scientists, discussed previously. More recently, several New Religious Movements, such as the Hare Krishna, The Unification Church ("Moonies"), the Divine Light Mission, and others, even though they were ephemeral, have created serious social and psychiatric problems because they have violated the civil liberties of certain cult members.

New Religious Movements also recruit emotionally disturbed youngsters, 30–38% of whom have been under psychiatric treatment before joining these cults [13]. These youngsters become easy prey of the cults because of the "affection and concern" shown by cult members, who are oriented toward communal or unconventional living, as a large extended family. Such cults, which exceed 2,500 in number in the US alone, provide emotional support, a mission in life, and a feeling of accomplishment that is badly needed by their members. On the other hand, some cults are a public health hazard, because their leaders have been found guilty of inflicting mental distress, coercive persuasion, peonage, unlawful imprisonment, sexual abuse, physical violence, deceptive fundraising practices, misuse of charitable status, and other abuses.

Concerned parents and friends of group members have occasionally counter-attacked by rescuing and "deprogramming" the recruits. Deprogramming consists in removing the cult members from the group, and subjecting them to a coercive reeducation process. This technique, even though successful in many instances, has created additional legal and psychological problems, because the rescue also violates the civil rights of the recruits. We cannot analyze here all the psychiatric and civil rights issues surrounding the cults and New Religious Movements, but the interested reader is referred to the American Psychiatric Association report [13] and to Galanter [14].

2.7 The tragic consequences of blind faith

There is obviously a wide spectrum of sects and cults. In the extreme cases, the cult leader is eccentric, with bizarre ideas, demanding absolute obedience. The leader becomes the self-appointed interpreter of God's will, and demands absolute faith, which may have tragic results. These leaders manage to convince the converted individuals to abandon their families, or bring them into the sect community, to which they must subsequently donate all their possessions. The followers often live in an unconventional manner under the guidance of an authoritarian and charismatic leader. In some cases, when the leader makes apocalyptic predictions such as the catastrophic end of the world with the salvation of the righteous, or the Second Coming of Jesus, they are prone to violence, which usually ends in tragedy when the predictions do not materialize. This is what happened with the mass murder–suicide of the People's Temple sect in Jonestown, Guyana in 1978, when more than 900 US citizens died.

Another incident in which a sect was involved, in 1993, destroyed the Mount Carmel Center, in Waco, Texas. There was an armed conflict between a religious sect, the Branch Davidians headed by David Koresh, and the Federal law enforcement agents, following a 7 week siege. The sect, which had a long history of apocalyptic predictions, separated from the Davidians Sect of the Seventh Day Adventists, who believe that Jesus' Second Coming and the end of the world are near. The first prediction, by William Miller (1782–1849), was that Christ was returning on October 23, 1844. When Jesus did not show up, this led to much disappointment [15]. Since then, the second coming has been supported by visions and delusional interpretations of the scriptures by Ellen G. White. She had about 300 visions in her lifetime, and seems to have been frankly paranoid: she thought that Satan, who was represented by the Bishop of Rome, would seek to control the world. She believed that only those whom God addresses audibly are able to understand and explain the Scriptures [15]. The most parsimonious explanation is that Ellen G. White was not only paranoid, but was also having very vivid auditory hallucinations.

In another apocalyptic prediction, Florence Houteff, the wife of a deceased leader of the Mount Carmel Center, announced that on April 22, 1959, the faithful would be slaughtered, then resurrected and carried to heaven on clouds. Nothing happened, so everybody was disappointed, and the prophet gradually lost her followers. It is clear that the Davidians have been waiting a long time for a bloody confrontation between the good forces of God and the evil intentions of Satan. The stage was set for the unfolding of the Waco tragedy.

The leader of Mount Carmel Center was David Koresh, a peculiar self-appointed Messiah. His interpretations of the scriptures are clearly delusional and self-serving. Biblical revelations prompted him to father many children by different concubines, several of whom were under age. In contrast, the males in the colony were required to abstain from sex. Moreover, he was known to practice cunnilingus, which was considered a diabolical act by some members of the sect. His sexual practices were an obstacle, not only for preserving the flock, but also for expanding their number. Koresh believed that he had to be, not only a lover but also a *fighter*, so he accumulated an arsenal of semi-automatic rifles, and materials to manufacture hand grenades and other explosives [15].

The recurrent apocalyptic prophesies of the sect were fulfilled when the Federal law enforcement officers, who were probably identified with the forces of Satan, prepared to invade the Mount Carmel complex. The final episode, on April 19, 1993, was brought about by a fire that consumed the Mount Carmel Center. It ended with the death of 76 men, women and children, including David Koresh. The origin of the fire that engulfed the Mount Carmel Center is in question and the Federal agents may share part of the blame. However, it is clear that such incidents do not develop unless faith is pushed to the limit, to create a shared psychotic disorder.

Independently of these events, another apocalyptic episode took place in late 1994, when 52 members of the "Order of the Solar Temple" were found dead in Switzerland and Quebec, in a murder–suicide ritual. Sixteen additional members of the same sect died in France, near the frontier with Switzerland, in December 1995. The bodies were found arranged in a sunburst pattern. Fourteen were shot after ingesting a sedative, and two—who were assumed to have shot the rest—died of self-inflicted gunshot wounds.

Sects whose beliefs generate violence or mass immolation attract attention, and they are labeled as deviant cults. The teachings and the doctrines of these charismatic groups are considered eccentric. In these extreme cases, it is clear that their leader is a disturbed paranoid individual who has induced a *shared psychotic disorder* in all the members of the sect. However, if the activities of the cult do not have tragic consequences, society does not recognize the psychotic character of the cult's teachings. The members of the cult could not be judged as delusional, because religious beliefs are protected by the American Constitution. The New Religious Movements are only judged by their results, and not by the delusional beliefs of their members. It is important to remember that not even the most grossly ridiculous religious beliefs are considered delusional because of a loophole in the official definition of delusion that is in compliance with the constitution [13].

Faith does not have any truthful cognitive value because it cannot distinguish objectively between true and false, and cannot distinguish religious extremism from moderation. Deductive and inductive logic, as well as empirical knowledge, have rules that preclude gross deviations from the truth. Despite its limitations, faith is considered by all religions as the highest form of knowledge. However, faith is the engine that has produced the most striking injustices in history, and keeps the faithful ignorant of what science has to say. Another of the self-serving theories that some theologians, preachers, and gurus have proclaimed is that religious experiences are the "highest forms of consciousness", which is a senseless metaphor in terms of truth. The mountains of incomprehensible gibberish created and propagated by religious writers hardly testify to the "superiority" of such states of consciousness. There is no acknowledgment of the fact that faith cannot discriminate truth, so any religion becomes true by definition. The tragedies generated by the cults, the ridiculous predictions of hundreds of visionaries and prophets, the Inquisitors of all times, the witch hunters, the religious persecutors and terrorists, are all based on "The True Faith". Indeed, religious faith is blind by necessity, because the "right" and the "wrong" faiths are not distinguishable by their truth value.

References

1. de Waal Malefijt A. Religion and Culture. An introduction to anthropology of religion. 1st. ed. London: Collier-Macmillan Limited; 1968.
2. Durkheim E. The Elementary Forms of Religious Life. New York: The Free Press; 1995.
3. Adler JM. Truth in Religion. The plurality of religions and the unity of truth. 1st. ed. New York: Macmillan Publishing Co.; 1990
4. Hofstadter R. Anti-intellectualism in American Life. 1st. ed. New York: Alfred A. Knopf; 1963.
5. Descartes R. Discourse on Method and The Meditations. London: Penguin Books; 1968.
6. Chalmers DJ. The Conscious Mind. 1st. ed. New York: Oxford University Press; 1996.
7. McGinn C. The Mysterious Flame: Conscious Minds in a Material World. First ed. New York: Basic Books; 1999.
8. Musacchio JM. The ineffability of qualia and the word-anchoring problem. Language Sciences 2005;27(4):403-435.
9. Snow CP. The Two Cultures. Cambridge: Cambridge University Press; 1993.
10. Dunlap K. Religion. Its functions in human life. 1st. ed. New York: McGraw-Hill Book Co., Inc.; 1946
11. Armstrong K. A History of God. First ed. New York: Alfred A. Knopf; 1993.
12. Kosmin BA, Lachman SP. One Nation Under God. Religion in Contemporary American Society. 1st. ed. New York: Harmony Books; 1993
13. Cults and New Religious Movements. A report of the American Psychiatric Association. 1st ed. Washington, DC: American Psychiatric Association, 1989.
14. Galanter M. Cults. Faith, Healing and Coercion. 1st. ed. New York: Oxford University Press; 1989.
15. Reavis DJ. The Ashes of Waco. An Investigation. 1st. ed. New York: Simon & Schuster; 1995

3 The evolution of human ancestors

Summary Fossil remains provide an incomplete, but objective glimpse into the evolution of our ancestors, which clearly indicate that *Homo* evolved from African hominids, but not from modern apes. For example, the most recent common ancestor of chimpanzees and hominids would have lived between 6 and 7 million years ago. However, human evolution is much more complex than previously suspected, because many hominid groups in fact evolved, but many also perished in the fierce competition for survival. This competition was won by *Homo sapiens*, perhaps as a consequence of his greater intelligence. *Homo erectus*, an ancestor of modern humans, left Africa about 1.8 million years ago. It was a bipedal hominid that walked upright and had a much smaller brain than our more recent ancestors. Early humans left a magnificent legacy of tombs that show their veneration of the dead. Because of their rudimentary knowledge, they relied mainly on supernatural (magical and spiritual) beliefs for explaining natural phenomena and for coping with their fears, diseases, and natural disasters. Thus, their supernatural and spiritual beliefs were partially incorporated into their religious beliefs and practices.

3.1 The veneration of the dead and early supernatural beliefs

There is no indication that animals other than humans conceive their own death, even though animals are hard wired to avoid pain and to defend themselves against predators. Some anthropoids, however, perceive the death of relatives as suggested by infants of the common chimpanzee (*Pan troglodytes*) that get depressed and lethargic at the death of their mother. Older siblings comfort the bewildered infants, who are sometimes adopted by an aunt or another adult. Despite the care and affection provided, the orphans still sometimes die within a few weeks or months of their mother [1]. Chimpanzee mothers are also known to continue carrying their dead babies in their arms for a few days, and elephants repeatedly return to sites where there are skeletons of their departed relatives, suggesting that they also mourn this loss.

J.M. Musacchio, *Contradictions: Neuroscience and Religion*,
Springer Praxis Books, Popular Science, DOI 10.1007/978-3-642-27198-4_3,
© Springer-Verlag Berlin Heidelberg 2012

According to the archeological record, early humans over 90,000–100,000 years ago started to care for the dead by burying them and setting them in special positions. There are also examples of buried children with the legs folded, and some sites contain little objects that were probably used as toys, such as small antlers. *Homo neanderthalensis*, a different species of hominid, unrelated to us (Fig. 3.1), also started to bury their dead about 65,000 years ago or perhaps even earlier, indicating that early humans venerated the dead. This practice later generalized in the most diverse cultures, some of which have left a magnificent legacy of tombs and a rich body of myths and legends.

In some cultures, the veneration of the dead progressed to the *cult of the ancestors*, which consists in the worship of or reverential homage to the spirits of dead relatives. The cult of the ancestors is also found as a component of more general systems of beliefs, but is not necessarily universal, even though it is evident in a variety of special rites and ceremonies. The cult of the ancestors presupposes the survival of the soul, and its capacity to interact with the living. The spirits of the dead are generally thought to help the living, but this is not always the case. Ancestor worship is widespread in Africa, Melanesia, China, Japan, India, and many Middle Eastern and European societies.

The myths and legends, in which different cultures believe, probably arose as much from human needs to predict the weather and the seasons as from questions about the Universe and the meaning of life and death. Myths have been defined in many ways, reflecting the points of view of the different professionals such as linguists, psychoanalysts, and anthropologists who made the definitions. Myths are stories with a plot, but they are highly symbolic, and have psychological, social, and religious implications. They can express anxieties triggered by natural or supernatural perils and can explain, in an allegoric fashion, the creation and origin of particular people. Myths are essential beliefs that codify and enforce the moral and social order of many early cultures. All human groups, from single tribes to ancient civilizations, had myths, or stories dealing with supernatural beings, ancestors, or heroes. These included exceptional individuals, such as chiefs or warriors. In primitive cultures, everything is also permeated by *animism*, which as previously discussed, is the belief that animals and objects have a soul or a spirit that is concerned with the survivors' affairs.

Mythical stories were used to answer questions about basic aspects of the group's existence, for instance: "Where do we come from? Who made us? How was the Earth created?" Such stories suggest that early humans probably shared our desires to know our origins and our ultimate fate, but they were not equipped to formulate their questions clearly, because of their lack of sophisticated language and conceptual structures. Besides, they had much more urgent and practical needs that may have overshadowed their desire to understand their origins and future existence.

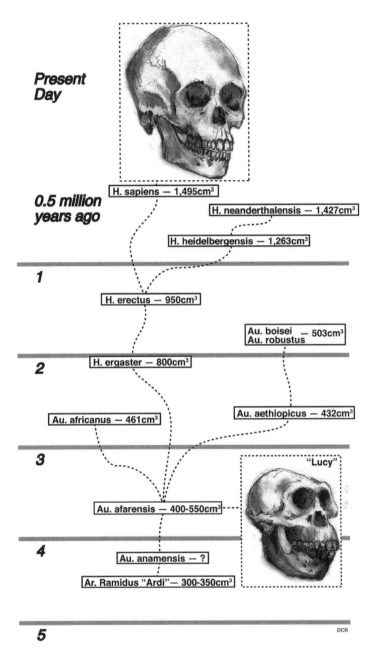

Present Day

H. sapiens — 1,495cm³

0.5 million years ago

H. neanderthalensis — 1,427cm³

H. heidelbergensis — 1,263cm³

1

H. erectus — 950cm³

Au. boisei — 503cm³
Au. robustus

2

H. ergaster — 800cm³

Au. africanus — 461cm³

Au. aethiopicus — 432cm³

3

"Lucy"

Au. afarensis — 400-550cm³

4

Au. anamensis — ?

Ar. Ramidus "Ardi" — 300-350cm³

DCR

5

Fig. 3.1   Evolution of Homo sapiens from African hominids. Ardipithecus ramidus —Ardi—is one of our earliest ancestors. He was arboreal, but walked upright. Ardi and other early hominids had projecting canine teeth typical of aggressive animals. In about a million years, she evolved into Australopithecus afarensis or "Lucy", before branching out into several hominids, which gave origin to Homo sapiens within the next 3 million years. The volume of their brains in cubic centimeters (cm³), as compiled by Allen [5], is included after their names. The dotted lines indicate the most likely branches of the evolutionary tree

Fear, much more than curiosity is what guided the actions of early humans: fear of hunger, sickness, and natural disasters. An Eskimo shaman told Rasmussen: "We do not believe: we fear." Fears of misfortune caused by sorcerers or spirits permeate the life of many primitive cultures [2]. To overcome their fears, early humans resorted to magic practices and rituals, in attempts to affect natural processes and find security. Magic is based on humans' confidence that they themselves can dominate nature directly. The anthropologist Bronislaw Malinowski (1884–1942) writes that the Sacred—Magic and Religion—dominated all early cultures. He maintained that customs and beliefs have specific social functions, that science in primitive cultures is rudimentary even today and is represented only by the arts and crafts invented for hunting, fishing, getting shelter, and meeting other practical needs such as weaving or making baskets. The successful practice of these abilities requires a keen knowledge of animals, plants, and natural processes. Despite some limited success, primitive science could not provide solutions to the countless problems that plagued early civilizations. Moreover, those civilizations could not control most natural processes, so early humans had little choice but to resort to supernatural practices, based on magic and rituals [3].

3.2 The variety of hominids and human ancestors

The evolutionary history of *Homo sapiens* is more complex than we ever suspected, so it is difficult to present the evolutionary tree of our ancestors, because the origin of its branches cannot be unequivocally identified or precisely dated (Fig. 3.1). Thus, only the most widely accepted chronological views will be presented here, bearing in mind that there are still some doubts about genetic relationships that will be settled when more details or DNA data become available. We know today that *Homo sapiens* had many distant cousins in Africa, because there are at least 10–12 different species of fossils of the genus *Homo* that have been described recently; they lived sometimes simultaneously, but in isolated groups, between 2 million and 30,000 years ago. Dating the individual fossils is difficult, and so also is establishing the relationships between them.

The oldest hominid remains that have some human features and walked upright were *Sahelanthropus tchadensis* and *Orrorin tugenensis*, which separated from the common chimpanzee's ancestors between 6 and 7 million years ago (Mya). These fossils are followed in the evolutionary tree by *Ardipithecus kadabba*, a precursor of *Ardipithecus ramidus* (Fig. 3.1). This was discovered in 1966 by Yohannes Haile-Selassie, at the beginning of his career [4]. *Ardipithecus ramidus* or "*Ardi*", also nicknamed "Root Ape", filled a gap that had remained empty for a while between *Ardipithecus kadabba* and *Australopithecus anamensis*, both discovered in Ethiopia. *Ardipithecus ramidus* is one of our earlier ancestors, dated to 4.4 million years ago (Fig. 3.1). This discovery by Gen Suwa and Tim White in 1992 *negated* the presumption that we are descended from African apes, since the latter evolved independently about 6.5 Mya and in a different direction. *Ardi* was partially arboreal but was able to walk upright; its brain had a volume between 300 and 350 cm^3, which is about *one quarter* the size of our brain [5]. Ardi and other early hominids had projecting canine teeth, typical of wolves and other aggressive animals. The transition to a more peaceful social structure of early hominids probably accounts for the progressive shortening of their canine teeth.

Another fossil that recently became the subject of many popular articles is Lucy or *Australopithecus afarensis*, which was discovered by Donald Johanson in 1974. Remains of this species were widely spread across Africa, where they lived between 3.1 and 4.4 million years ago. They are a closer predecessor of *Homo* than Ardi (Fig. 3.1). Lucy is a bipedal fossil hominid that walked upright, was terrestrial, and had a larger brain than *Ardipithecus*, in fact 400–550 cm^3, a volume comparable with living chimpanzees [6]. These and other estimates of the brain size of our early ancestors are possible by the direct correlation existing between the volume of the internal cavity of the skull and the weight of the brain. Lucy, *Australopithecus afarensis*, is considered an ancestor of *Homo habilis* (the handyman), who evolved more than 2 Mya and could have been an ancestor of *Homo erectus*. As indicated, the latter was one of our closest ancestors (Fig. 3.1). There were several other species of the genus *Australopithecus* which had a small brain compared with their contemporary hominids, and became extinct between 1.5 and 4 Mya.

What is interesting about the African hominids is that, if they had only been arboreal, their mobility would have been quite limited, whereas they were in fact bipedal and could thus walk upright and move around a much larger area, important for obtaining food and water. In any case, it is reasonable to think that, for safety or family reasons, their descendants would have stayed for many generations near the place in which they were born. The groups were probably small due to high infant mortality, so they would have had limited inter-breeding. Thus, cumulative random mutations over several thousand years would have added to their isolation from other groups, fostering the potential to create new species. At least 90% of our genome comes from African ancestors, but these estimates may change slightly with the development of better techniques to analyze DNA sequences. These sequences are often hard to determine accurately, because most remains have been contaminated with DNA from modern humans, animals, and bacteria [7].

Homo erectus, a genetically diverse ancestor of modern humans arose in Africa and migrated to Europe and Asia as early as 1.8 Mya (Fig. 3.1). The new environment and new challenges most likely put *Homo erectus* to the test, because at the time Homo *erectus* had a cranial capacity of only about 950 cm^3, which is just 63% of the volume of our brain. This means that *Homo erectus* was not brilliant, but was nevertheless determined to find a better environment. In Asia, *Homo erectus* interbred with *Homo neanderthalensis*, a different species who lived in Europe from 350,000 to 30,000 years ago and who had previously diverged from *Homo heidelbergensis*. When *Homo erectus* left Africa, *Homo neanderthalensis* was in transit from Western Europe to central Asia, but contrary to previous impressions, modern humans do not possess Neanderthal DNA [8]. The Denisovans, another group of hominids found in the Denisova Cave in the Altai Mountains in southern Siberia, have Neanderthal but not modern human DNA. Denisovans share 4–6% of their DNA with Melanesians from Papua and New Guinea [7].

Despite the multitude of relatives of the genus *Homo* that existed in Africa, the only species that remains alive today is *Homo sapiens*. It is composed of several closely related races, adapted to local environmental conditions. Adaptation explains why Scandinavians and Northern Europeans, who have lost their skin pigments, survive better than black people in northern Europe, since they have evolved to be better able to synthesize vitamin D, while black people, whose ancestors remained in Africa, have retained their skin coloration as a protection against the tropical sun. The different races living today are most likely the result of local mutations and also the isolation in which some of our ancestors lived for many generations. However, changes in future lifestyle, with much higher mobility and more frequent interracial unions, will surely contribute to creating a uniform population. This would actually slow down human evolution for the next few millennia. Eventually, humans may decide that they are ready to delete some of the genes responsible for diseases and aggressiveness, and to introduce genes that will increase intelligence and lifespan.

3.3 Human language was also born in Africa

The phonetic diversity of southwest Africa is the richest in "click" sounds, which are very ancient signals and suggests that not only the genus *Homo* but also language was born in Africa. This is supported by the parallel decrease in both genetic diversity of human populations and in phonetic variety of click sounds as early *Homo sapiens* moved away from western South Africa. This is statistically based on the decreased number of phonemes (simplest distinguishable sound elements) that languages use when small populations move away from where they originated [9]. There is also general agreement that language provided humans with a secret and powerful weapon to plot against animals and other hominids who did not understand their language.

3.4 The evolution of the brain: from hominids to *Homo sapiens*

Our knowledge of the hominid brain progressed slowly to begin with, but new discoveries present a complex picture. Initially, the evolution in the size of the brain from 6 to 2 Mya was relatively slow and remarkably stable, being around 400–450 cm^3, which is comparable to that of the great apes [5]. About 2 Mya, the hominid brain was slightly larger than ape brains, and intelligence started to play a major role in survival and natural selection, especially with the development of the first stone tools. From 1 to 1.5 Mya, the brain expanded rapidly to about two thirds of its present size (Fig. 3.1). In other words, early hominid brains started to grow much faster and they have almost quadrupled in volume since then. This greatly increased the probabilities of survival through the development of the brain regions in the frontal lobes controlling language and creative intelligence [10, 11]. With the progressive development of higher intelligence, abstract thinking, and language, humans were in a better condition to face life and potential enemies. With the invention of agriculture, which spread about 10,000 years ago, there was some free time between harvests to invent new tools, develop new ideas, and wonder about the meaning of life and death. Thus, culture evolved in parallel with the religions and early technological achievements of our human ancestors.

Considerable insight into the human brain comes from observing it during its pre- and postnatal development. The brain centers that serve the most basic functions, such as breathing, suckling, and reacting to pain, develop first in the individual (ontogeny) and in the species (phylogeny), whereas the more recently evolved parts of the brain take longer to mature. Thus, even though teenagers look fully developed, their brains are not mature until their early or mid-twenties. Therefore, it is not surprising that most countries recruit soldiers before they reach maturity and could question the wisdom of war, or become conscientious objectors.

The prefrontal area of the modern human brain (Figs. 3.2 and 4.1) constitutes about 29% of the cerebral cortex, whereas this region comprises only 17% in the chimpanzee, 11.5% in the gibbon, 7% in the dog, and only 3% in the cat [10]. In addition, the human brain cortex has much more complex neural networks and its brain cells are fully packed and heavily interconnected, not only within the frontal cortex but also with the rest of the brain. The surface of the human brain is actually much larger than it looks, because it has many wrinkles and folds, which help to accommodate more neurons per unit surface. The rich circumvolutions that characterize the human brain are absent in lower animals, like fishes, amphibians, and reptiles. The latter have a smooth brain surface and for this reason are called *lissencephalic* animals.

As can be observed in Fig. 3.2, the rat and the mouse also have brains in which circumvolutions are hardly noticeable. The more evolved the brains are, the thicker and more convoluted the brain cortex is, because it contains more neurons, fibers, and synaptic contacts per unit surface. The density of neuronal packing is what makes the wrinkles in the brain of the more intelligent animals.

The human frontal lobes, especially in the convexity of the prefrontal area, are the places where plans are made and behavior is organized according to the goals of the individual. This requires orchestrating the activity of the rest of the brain while keeping in mind the overall purpose of the plan. The frontal lobes

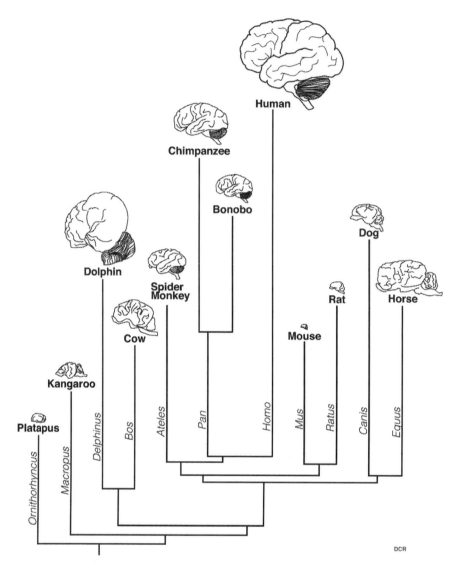

Fig. 3.2 Brains of different animals of the Class mammalia. The drawing approximately illustrate their relative size. The branching lines indicate their evolutionary relationships into order, family, subfamily, genus, species, subspecies, etc. For the sake of simplicity, only the Genus is indicated on the vertical lines. Only the most common dolphin is listed, because the dolphins have many genuses

also have an inhibitory role that is useful when the predicted actions are thought to have negative consequences for the individual or relatives. Thus, the frontal lobes control and coordinate all cognitive functions to achieve specific goals. The left-lateral inferior prefrontal cortex is directly involved in processing and integrating language and organizing actions in the temporal dimension. This is why it is sometimes said that the frontal lobes are in charge of the *executive functions* of the individual. In addition, the left frontal lobe near the traditional area of Broca (Fig. 4.2) is instrumental in the temporal organization of language [10]. The frontal lobes, in association with the left temporal lobe, are crucial in determining the semantic and symbolic role of language, which is perhaps the most distinctive human attribute.

The importance of the prefrontal areas for abstract thinking, creative intelligence, and moral judgment has been demonstrated beyond any doubt by the study of a large number of soldiers that recovered from head wounds after the Second World War, by the effects of frontal leucotomy, and by the side-effects of removing tumors from the frontal poles. The phylogenetic development of the prefrontal cortex is directly related to higher intelligence, abstract thinking, problem solving, and language [10, 11]. The more evolved frontal lobes enabled our ancestors to start developing scientific explanations obtained by creating mental models that actually mimic natural processes. The capacity to create mental models that fit, predict, and explain reality is the product of the empirical method, which relies heavily on observation, experimentation, reason, and logic.

References

1. Goodall J. Understanding chimpanzees. 1st ed. Cambridge, MA: Harvard University Press; 1989.
2. Lèvy-Bruhl L. Primitives and the supernatural. First ed. New York: E.P. Dutton & Co., Inc.; 1935.
3. Malinowski B. Magic, Science and Religion and other essays. First ed. Boston, MA: Beacon Press; 1948.
4. Gibbons A. The First Human. First ed. New York: Anchor Books; 2006.
5. Allen JS. The Lives of the Brain. First ed. Cambridge, MA: The Belknap Press; 2009.
6. Gibbons A (2006) Lucy's Tour Abroad sparks Protests. Science 314, 574–575. 10–27-2006.
7. Gibbons A. A New View of the Birth of Homo sapiens. Science 331, 392–394. 1-28-2011.
8. Pennisi E. Tales of a Prehistoric Human Genome. Science 323, 866–869. 2-13-2009.
9. Atkinson QD. Phonemic diversity supports a serial founder effect model of language expansion from Africa. Science 2011;332(6027):346–349.
10. Fuster JM. The Prefrontal Cortex. Fourth ed. London, UK: Academic Press; 2011.
11. Deacon TW. The Symbolic Species: the co-evolution of language and the brain. First ed. New York, NY: W. W. Norton & Company, Inc.; 1997.

4 The most amazing window on human nature

Summary Brain diseases are the most cruel natural processes, but they provide an objective view of human nature that reveals the fascinating complexity of our minds. Abnormal electrical activity of the brain can produce sensory or motor epilepsy, or affect areas that produce deep psychological and ecstatic experiences. Neuroscience provides evidence that the mind does not consist in "psychological" processes, but in *physical* events that take place in the brain. The disintegration of the self produced by several diseases indicates that the Cartesian idea of the self, a physical body with a soul, which was conceived as a homogeneous supernatural substance, is unjustified. The success of psychoactive drugs in treating emotional and behavioral abnormalities also shows that emotions and behavior are physical processes.

4.1 María's brain was burning her hand

María was probably 15 years old when she came to the epilepsy clinic with her mother, who explained that she was having "attacks." María would suddenly pass out and develop generalized convulsions causing her to fall and bite her tongue. These episodes occurred several times a week and seriously disturbed her schoolwork and social life. Sometimes, a brief feeling of pins and needles and a burning sensation in her right hand and forearm, which is called the aura, "the breeze that announces the storm", preceded the spells. The neurologist diagnosed that María had epilepsy and prescribed phenytoin, which was the most effective anticonvulsant for the treatment of grand mal epilepsy in the 1950s.

María returned to the clinic 2 weeks later, but she was very upset and agitated, "because you made me worse: now I have terrible attacks of burning in my hand and arm!" (Fig. 4.1).

Actually, she had fewer convulsions, so her mother was happier, but María was not, because she was able to perceive the unpleasant burning spells in her

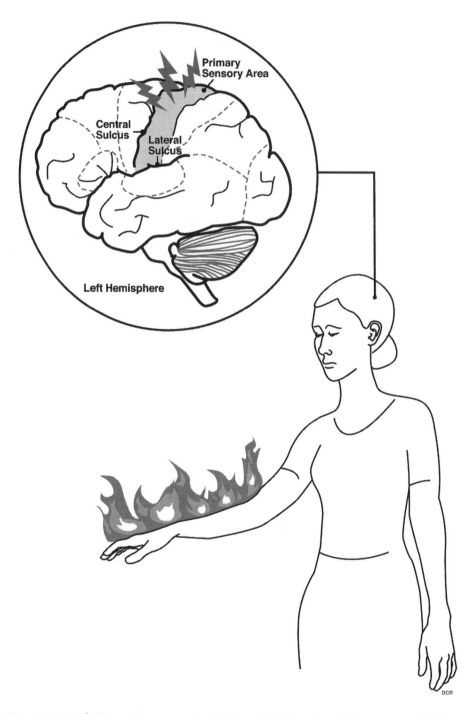

Fig. 4.1 María's burning sensation in her right hand and forearm. María's burning sensation actually originated in the left hemisphere of her brain (enlarged in circle). Also shown is her abnormal electroencephalographic activity ("lightning bolts") in the left primary sensory area of her brain

right hand and forearm, which were no longer masked by an immediate loss of consciousness and a general convulsion. As medical students, we were surprised and disappointed, but the neurologist explained that the dose of the medication was not high enough to suppress the initial epileptic activity, even though it was enough to block the spreading of the convulsion to the rest of the brain. María was better because she did not pass out so frequently, but she felt worse because she remained alert to feel the epileptic discharges produced by her left brain, which would normally receive sensations from her right hand and forearm (Fig. 4.1, top). The anticonvulsant drug also stopped the spreading of the abnormal electrical activity to the brain regions that maintain consciousness, so she did not pass out. The neurologist increased the dose of phenytoin, and she was much happier on her next visit because her convulsions were now sporadic and the burning sensation was mild.

We were amazed that the epileptic discharge, an abnormal electrical activity in a sensory area of the *left* brain, felt as if it was a *real* burning sensation in her *right* hand. However, María was also sure that her spells originated in her right-hand and forearm, despite the fact that her disease was actually in the primary sensory area of her left hemisphere (Figs. 4.1, 4.2).

The neurologist explained that the *feeling in her right hand and forearm* was produced by a sensory epileptic fit in her left-brain. María probably had a brain scar involving the area in the sensory cortex that normally receives information from the hand and arm. It was a revelation to the young trainees, as indeed it was to John Hughlings Jackson, the first neurologist to describe in 1876 that *what happens in the brain is what determines what we feel* [1]. This means that we do not perceive sensations with our hands, where the external sensors are located; we only perceive the brain activity produced by the *signals* that originate in the peripheral body. It also means that when we touch an object, we do not *feel* the object with our hands. Instead we perceive the *encoded message* that travels from the hand touch receptors, through the nerves in the arm, to the brain, where the message is interpreted and perceived. As early as the eighteenth century, the idealist philosopher George Berkeley (1685–1753) had a point in thinking that we perceive only ideas or experiences, but not matter [2]. However, he wrongly denied the physical existence of external objects and matter, and failed to realize that ideas were in fact physical processes that take place in the brain.

4.1.1 Epilepsy shows that sensations and movements are encoded in the brain

María had *Jacksonian* epilepsy which originated in the left-brain sensory area of the hand and arm, and which later spread to the rest of the brain and produced a full epileptic convulsion with loss of consciousness. The term Jacksonian refers to John Hughlings Jackson (1835–1911), a neurologist practicing in

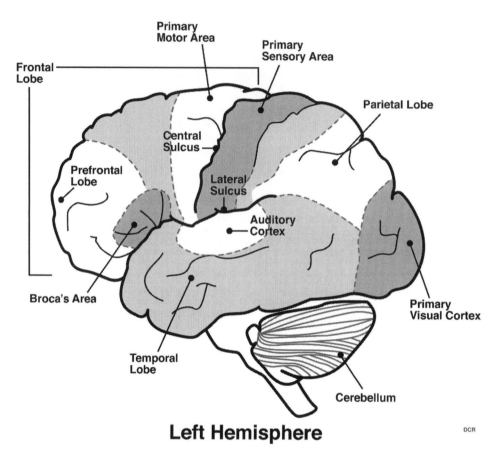

Primary
Motor Area

Primary
Sensory Area

Frontal
Lobe

Parietal Lobe

Central
Sulcus

Prefrontal
Lobe

Lateral
Sulcus

Auditory
Cortex

Broca's Area

Primary
Visual Cortex

Temporal
Lobe

Cerebellum

Left Hemisphere

DCR

Fig. 4.2 Lateral view of the left hemisphere of a normal brain. It shows the primary sensory area where María's convulsions originated. Other areas shown include Broca's area (responsible for language articulation), the temporal lobe auditory cortex, and the external parts of the primary visual areas. For additional views and explanations of the visual areas, see Chap. 5 and Fig. 5.3

London from 1860 to the beginning of the twentieth century. He was fascinated by the information that could be extracted by simple observation, and by carefully questioning of patients and their relatives. He correlated the symptoms observed with the localization of brain lesions, and postulated the "representation"[1] of movements and sensations in the contralateral brain cortex (Fig. 4.3).

---

[1] Representation in this case refers to the brain areas where the peripheral signals arrive and from where the electrical stimulation of the brain cortex can produce sensations in specific areas of the body.

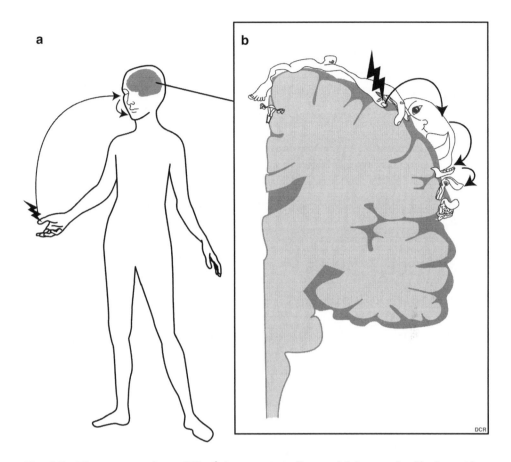

Fig. 4.3   The progression of María's sensory epilepsy. (a) As perceived by her and as represented in the primary sensory area of her brain (**b**). In (**a**) the arrows indicate a tingling sensation which María's felt under the nail of her right thumb (lightning bolt), and which later "jumped" to the right side of her upper lip, face, and tongue. In (**b**), the arrows show the start of the convulsive activity in María's left brain (lightning bolt), which produces a tingling sensation in her right thumb and later spreads to her face and body. The drawing of the sensory homunculus is based on studies of Penfield, in which the most sensitive body parts (e.g., lips and mouth), occupy larger areas than the less sensitive regions such as the leg and the foot of the sensory homunculus

Jackson also observed that epileptic fits involving the right side of the body, frequently produced a transient loss of speech. This is produced by epileptic activity on the speech centers which are located on the left side of the brain in most right-handed people. This finding was in agreement with earlier reports by Pierre Paul Broca and others, who discovered that speech control is localized in the left side of the brain, in what was later called Broca's area (Fig. 4.2).

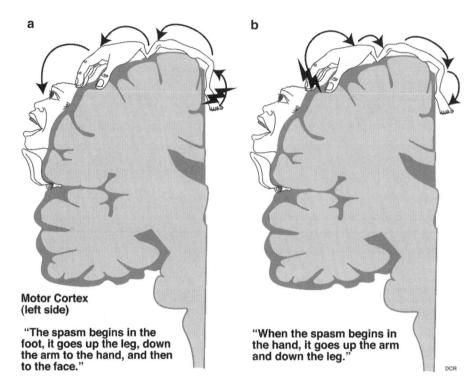

a

b

Motor Cortex
(left side)

"The spasm begins in the
foot, it goes up the leg, down
the arm to the hand, and then
to the face."

"When the spasm begins in
the hand, it goes up the arm
and down the leg."

DCR

The Jacksonian March

Fig. 4.4 Jacksonian March. Diagrammatic representation of the "Jacksonian March" through body parts in the primary motor cortex. The arrows in (**a**) depict Jackson's observation in his own words:"when the spasm (lightning bolt symbol) begins in the foot, it goes up the leg, down the arm to the hand, and then to the face". In (**b**), the arrows indicate, "When the spasm begins in the hand, it goes up the arm and down the leg"

These observations established that the site in the brain where the convulsion starts would determine the clinical characteristics of the epileptic seizure. Thus, if a convulsion does not spread rapidly enough for patients to lose consciousness immediately, the patients can witness the effects of the convulsion and perceive the localized involuntary movements, loss of speech, or illusory sensory symptoms such as tingling or burning in the corresponding areas of their body.

Jackson recognized *sensory* epilepsy as early as 1862. He described the case of a woman whose seizures started with a tingling sensation in the right thumb, under the nail, which extended to the right side of the upper lip, face and tongue (Fig. 4.3). In many cases of motor epilepsy, the seizures spread in a fixed order following the motor representation in the brain cortex. Jackson observed that "when the spasm begins in the foot (Fig. 4.4a), it goes up the leg, down the arm to the hand, and then to the face, and when it begins in the hand (Fig. 4.4b), it goes up the arm and down the leg." The progression of the convulsion following up or down the *representation* of movements in the motor area was described by Jackson as "the March of the fit", but since then it has been known as the *Jacksonian March* (Fig. 4.4).

The motor representation of the body in the brain (Fig. 4.4a) helps to explain why a seizure travels between the hand and the face without passing though the arm and forearm. This clearly establishes that the spreading of the seizure takes place in the brain cortex, where the areas of the face and hand are nearby, whereas in the periphery the whole arm separates the face from the hand. Jackson said in 1875, "Paralysis and convulsions are the results of experiments made by disease on the brain of man" [1].

4.1.2 Consciousness and the mind are functions of the brain

John Hughlings Jackson and many neurologists after him have described a multitude of *psychological* symptoms produced by epilepsy, specifically those involving the temporal lobe. Localized temporal lobe (Fig. 4.2) seizures produce a large variety of complex psychological, sensory, and motor symptoms, including the patients feeling that their spirits are separating from their bodies, as shown in Figs. 1.1 and 1.2. When the abnormal electrical activity spreads to the whole brain, the patient loses consciousness and develops generalized convulsions. Briefly, psychic seizures include illusions, emotions, hallucinations, and forced thinking. Illusions can be visual, auditory, gustatory, etc., or produce a feeling of familiarity (déjà vu), remoteness, or strangeness. Emotional symptoms may include fear, loneliness, or sadness. Hallucinatory seizures resemble vivid dreams and can be as complex as a real life visual scene with voices and music. The forced thoughts are *intellectual auras*, thoughts of stereotyped pattern or crowding thoughts that the patient cannot describe clearly.

Jackson was fascinated by what he called "an experiment made on the brain by disease." He clearly stated that paralysis and convulsions supply evidence about the localization of movements and sensations in the brain. Using his method, Jackson was very successful—by the standards of the day—in localizing brain lesions. He proposed that, inside the brain, there are representations of impressions and movements in time and space. Jackson concluded that the brain, even though the organ of consciousness, was subjected to the same laws as the lower neural structures. Clearly, his ideas did not allow any exemption for the "spiritual" functions of the central nervous system (CNS). Thus, more than a century ago, consciousness was already known to be a physiological process produced by the brain, and not a manifestation of any supernatural spiritual entity [3].

Reports of deep psychological experiences produced by epileptic attacks are as old as history, but they have never been so vividly described as by Fyodor Dostoyevsky (1821–1881), the Russian writer famous for the profundity of his description of psychological states and religious mysticism. Dostoyevsky's epileptic attacks started with an ecstatic aura that he described in his "Letters," and through several of his fictional characters. In *The Possessed*, Kirilov says, "he suddenly attained the presence of eternal harmony" [4]. The feeling was so terribly clear and intense, that he had the impression "to live a lifetime in 5 s". He compares these experiences with those of Mohammed's visions of Heaven, which also occurred in a flash [5].

Delusions of detachment of the spirit from the body are frequent experiences during ecstatic auras. They seem similar to the mystical experiences reported by St. Theresa and other visionaries. The most important characteristic of these delusions is that the epileptic patients *are absolutely convinced that their souls had temporarily left their bodies, and actually visited Heaven, or the Land of the Spirits* [4]. Interestingly, epilepsy has also been known as the *Sacred Disease*, because epileptics were thought to be possessed by a supernatural being that induced periodic ecstasies and prophetic trances. Thus, the historic, literary, and medical evidence indicates that epilepsy can produce *delusions of supernatural experiences*, which deeply affect what is considered to be the soul. These experiences are similar to near-death experiences and to the hallucinogenic effect of some drugs, such as psilocybin.

Jackson stated in 1868 that "the psychological, like the physical processes of the nervous system, can only be functions of complex combinations of motor and sensory nerves." He realized that movements, sensations, and complex behavioral actions produced by epileptic fits resulted from "discharges" of the cortex, whereas paralysis resulted from destructive lesions. The fact that epileptic discharges produce movements, sensations, emotions, and automatic behavior shows that the brain mechanisms that underlie these diverse processes are all of identical nature.

In the preface to his paper on the localization of movements in the brain, Jackson wrote in 1876 that "the brain, although the organ of consciousness, is subjected to the laws of reflex action; and that in this respect it does not differ from other ganglia of the nervous system." He thought that during and after epileptic attacks, consciousness could be altered in various degrees, as happens with all other brain functions. He stated, "There is no such entity as consciousness; in health, we are from moment to moment differently conscious." He added, "Our present consciousness is our now mental state." What Jackson implied is that consciousness is a *functional* property of the brain, not a supernatural entity. "Consciousness varies in kind and degree according as the parts of the brain activity are different." ... "The explanation here combated is a metaphysical one." He was aware that consciousness was determined by the function of the different brain areas. Consciousness was the product of the "highest centers," but he admitted that his theory was imperfect, and indicated: "We do not really suppose there to be one fixed seat of consciousness".

Jackson was delighted when neurophysiologists established in animal studies that the electrical stimulation of the brain cortex produced movements on the opposite side of the body [1], just as he had predicted by observing his patients. These experiments confirmed his views that the "irritation" of specific brain areas produced the symptoms that he observed in his epileptic patients. Moreover, his findings supported the idea that apparently purely *psychological* phenomena could be triggered by the *physical* stimulation of the brain.

Epilepsy, a disease produced by bursts of abnormal electrical activity of brain cells, provides another clear example of the interactions of physical phenomena with what is generally considered a psychological or spiritual reality. Moreover, the marked similarities of the trance-like states described by many prophets and visionaries with ecstatic experiences, suggest the delusional origin of prophetic visions, revelations, and near-death experiences in which the spirit seems to abandon the body (Fig. 1.2).

4.1.3 "It must be great fun to speak to the preparation"

Jackson's general ideas on the representation of sensations and movements in the brain cortex were dramatically confirmed and expanded by the studies of Wilder Penfield (1891–1976), a pioneering neurosurgeon who operated to remove epileptic focuses in the brain of patients not responding to other treatments. Penfield took advantage of the curious fact that *the brain itself does not have pain receptors,*[2] so it does not hurt when touched, electrically stimulated, or cut. Thus, Penfield used local anesthesia only to open the scalp and penetrate the skull; he then stimulated electrically the brain surface of conscious patients to map different functions. These procedures were necessary to locate the epileptic focus and to avoid areas such as the language centers that obviously should not be removed. Electrical stimulation of language related areas is still used during brain surgery. Penfield drew detailed brain maps of the sensory (Fig. 4.3b) and motor (Fig. 4.4) representation of the body. The patients were awake, so they reported their experiences in detail. Stimulation of the sensory cortex produced sensations similar to those described by Jackson's patients, such as numbness, tingling, and burning in the corresponding parts of the opposite side of the body. Similarly, the electrical stimulation of the

---

[2] The inability to sense itself is perhaps the most misleading characteristic of the brain. This creates the pervasive illusion that all mental activities, including qualitative experiences, are mysterious non-physical processes, and fuels the idea of the supernatural soul (see Chap. 5).

primary motor cortex produced movements in specific groups of muscles, as previously reported by physiologists working with small animals and monkeys. Penfield also made motor maps (Fig. 4.4) that were similar to the maps previously obtained in chimpanzees.

The stimulation of other areas of the brain evoked specific memories and feelings that indicated the *physical nature of what is commonly called "psychological"*. Thus, Penfield's studies were extremely important because they not only confirmed many of the predictions of Jackson and others, but also demonstrated the physical nature of experiences that were previously thought to be of "spiritual" character. In addition, the electrical stimulation of the temporal lobe cortex can produce specific recollections, and complex visual as well as auditory hallucinations. Experiences produced by the brain's electrical stimulation are so vivid that one of Penfield's patients argued that she was actually listening to a record player in the operating room. Some of the patients experiences during brain stimulation were reported as "more real than memories", because they also included detailed background noises in addition to specific conversations and visual components. These sensations also included the emotions that the patients had at the time. The production of subjective experiences by the electrical stimulation of the brain is comparable to the production of muscle contractions induced in frogs by Volta and Galvani in the 1790s [3]. In both cases, the electrical stimulation of the tissues produced effects that were previously attributed to supernatural spirits.

The electrical stimulation of the brain was perfected in the course of many animal studies. One of the pioneers in the development of the methodology was Penfield's former teacher, Sir Charles Scott Sherrington (1857–1952), a British physiologist who received the 1932 Nobel Prize in Medicine for major advances in the understanding of the nervous and muscular systems. Sherrington had previously performed many experiments, similar to those of Penfield, but using animals. When Sherrington was 90 years old and retired, Penfield went to visit him. He was delighted to hear about Penfield's experiments, and said: "It must be great fun *to speak to the preparation*". What Sherrington had in mind was that, in the animal experiments that he pioneered, he could only guess through behavioral clues what the animals were feeling. He obviously understood the advantages of talking to the experimental subjects.

Penfield showed that the electrical stimulation of the brain cortex could revive deep psychological experiences, with all their visual and auditory details and with the original emotional components. He showed that the *brain stores experiences in a physical form*. When these records are replayed by electrical stimulation, they can occupy the focus of consciousness as can any other experience. This means that memories and the associated emotions are encoded in specific neuronal circuits. The details of memory encoding are not completely understood, but there is no doubt that the mechanisms involved are of a physical nature. Brain imaging techniques have confirmed and greatly expanded our ideas about the representation of sensations and movements in the brain cortex, a concept that Jackson pioneered.

4.2 The missing arm that hurts

My initial encounters with neurological patients were as fascinating as they were depressing. The fascination was produced by the curious symptoms of brain lesions and by what they reveal about the complexities of human nature. The depressing aspects arose from seeing the loss of memory, motor, and other essential functions in patients with strokes, amyotrophic lateral sclerosis (ALS or Lou Gehrig's disease), or Alzheimer's disease. Despite my inability to cure many of them, I was usually able to make patients more comfortable. Thus, my curiosity and the desire to help overruled my negative feelings.

Ricardo, a patient that I examined as a medical student, was strangely similar to María. He was a middle-aged retired construction worker, who went to the clinic complaining of pain and a burning sensation in his right arm. The problem was that *his right arm was missing*: he had lost it years earlier in an accident. Pain in the missing arm! How can an arm that you do not possess hurt? Ricardo pointed in the space where the *imaginary arm* hurt. The arm was missing from his body but not from his brain. The mental image of his arm was slightly smaller than the arm that he lost. One of the medical students thought that Ricardo was hysterical, but the neurologist in charge knew better and said that it was a classic case of pain in the *phantom limb*. Another student asked, "What is a phantom limb?" The neurologist explained that it is an illusion always present immediately after an amputation. The amputated limb is felt as still present, to the point that some patients have fallen out of bed when they tried to use their missing leg. The body image recorded in the brain needs new experiences to change. This suggests that we are not our physical body, but the *mental image of our body*, which the brain has physically encoded and stored from previous experiences.

These clinical observations show that what we perceive is determined by the nerve signals that arrive at the appropriate areas of the brain. Regardless of the actual source of the nerve signals, the pain is attributed to the peripheral territory where sensations are assumed to originate. For example, when accidentally hitting the elbow on the "funny" bone, we are in reality hitting the ulnar nerve that sends the "funny" message to the brain by producing a neural "short circuit". The pain and the pins and needles seem to come from the peripheral territory of the nerve: the pinkie and annular finger and the internal (ulnar) side of the hand and forearm, but they are actually produced by the mechanical stimulation of the ulnar nerve at the elbow.

The shrinking of the phantom limb observed after an amputation in humans is paralleled by the shrinkage of the brain representation of the fingers of a monkey after cutting their sensory nerves. These two observations are related because the image of the body stored in the brain is maintained by the incoming sensory messages. When the messages are discontinued, the brain image of the isolated or amputated part slowly shrinks, because the brain region is invaded by nerve fibers and messages from other areas, such as the face. However, if the scar in the stump irritates the severed nerves and produces pain, the phantom limb tends to persist. The changes in the brain sensory areas and in the subjective perception of the arm illustrate that *the brain is plastic and is in a constant state of reorganization*. The plasticity of the brain is one of its most remarkable characteristics and illustrates the tight relationship between form and function. These observations have been repeatedly confirmed in many animal experiments and clinical cases. Plasticity is one of the main properties of brain nerve cells and networks, and it is the very thing that makes learning possible.

4.3 "There is a stranger in my bed"

The integrity of the physical body mental image seems to be ingrained in the self, as happens with amputees who continue "feeling" and sometimes trying to use their absent limb. In contrast to lesions of the body, damage of higher brain centers can produce a phenomenon that we conceptualize as the reverse of a phantom limb: the patients have an intact body, but the mental map of their bodies is altered. We believe that we perceive our body directly, as it is. However, our body is only known by constructing a mental image that actually consists of physical changes encoded in cellular brain networks. The fact that brain lesions, hallucinogenic drugs, or emotional states can alter our body image indicates that our image is different from the body itself. This implies that our corporeal awareness is not innate. We have unconsciously collected the impressions of our own body and stored them as our body image.

Macdonald Critchley distinguishes the *conceptual* from the *perceptual* aspects of the body image, which he calls *corporeal awareness* [6]. Tactile, visual, and positional awareness contribute to our body perception and to the development of our mental image. For example, Critchley indicates that congenitally blind children who have never seen their bodies tend to exaggerate the relative size of their hands, lips, and mouth when making clay models. This distortion is most likely related to the primary importance assigned to the parts of the body with which they mostly sense the external environment.

The conceptual and perceptual aspects of our corporeal awareness develop during childhood and adolescence and continue to change throughout life. Some aspects of our body image change not only with our emotional state, but also with the social circumstances and attires. The multitude of factors necessary to develop the body image indicate that the process is complex and could potentially be distorted in a variety of fashions by lesions of the corresponding brain areas. The disorders of the body image are determined in part by the location of the lesion and are affected by the alertness of the patient. We mostly know about disorders of the body image associated with lesions of the non-dominant side of the brain, because lesions of the dominant side (usually the left) are accompanied by serious language impairment.

One of the mildest disorders of the body image is called *unilateral neglect*, because the patients seem to "forget" to use spontaneously the limbs on one side of the body, even though the limbs can be moved on specific command. This neglect can also be referred to the tactile space, as commonly shown in patients who dress only one-half of their body, or shave only one half of their face. In the most pronounced cases, the patient may feel that one-half of his body has disappeared. Under the effects of LSD, one of my patients with a parietal lobe lesion said, "I feel like a chicken cut in half" [7].

Disorders of the body image are also complicated by the attitude of the patient toward the disease, by their alertness, and by the extent of the brain damage, which may impair the patient's collaboration with a complex examination. Thus, the symptoms may vary from a mild lack of concern to a complete denial of the illness, with elaborate stories (confabulations) to cover up the symptoms [8]. For example, when the patients walk, they may have the tendency to bump into objects located on one side. Neglect can be observed on either side, but right-side neglect is more difficult to study because it is produced by left-brain lesions that are frequently accompanied by speech disorders. Patients frequently say that the insensitive or paralyzed limb, which they can touch and feel with their healthy arm, *belongs to a stranger.* At other times the patient may insist: "There is a stranger in my bed." The "stranger" is sometimes given a name and can be considered friendly, but may also produce erotic feelings, bother, or sexually assault the patient.

Another disorder of the body image is the *alien hand syndrome*, in which the affected limb may move involuntarily, hitting or choking the patient who tries with varying degrees of success to hold and restrain the arm. Patients often personalize the arm, indicating that it belongs to a stranger or treating it as a misbehaving child [9, 10]. The arm also may perform simple involuntary activities, such as grasping objects, opening faucets, using tools, or making symmetrical movements. The alien hand syndrome is associated with a variety of brain lesions.[3] However, the extreme cases of distorted body image are

---

[3] The alien hand syndrome is associated with two main locations of the lesion. (1) The anterior or motor syndrome is produced by infarction or hemorrhage in the territory of the anterior cerebral arteries, which result in damage of the anterior corpus callosum and/or anteromedial frontal cortex. (2) The posterior or sensory form is due to corticobasal degeneration or posterior cerebral artery occlusion [9–11].

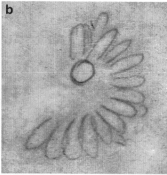

Fig. 4.5 Left visual sensory neglect. The visual neglect became evident when the patient was asked to draw a flower with a model in view (**a**) and from memory without a model (**b**)

mostly associated with large lesions that produce serious cognitive disorders and are also affected by patient alertness. Kurt Goldstein (1878–1965) remarked several times that the symptoms presented by a patient are quite complex, because they are not only the result of the disrupted function, but also of the *reaction and accommodation* to the catastrophic effects of the disease [12]. He also said: "Nobody will doubt that the observation and analysis of pathological phenomena often yield greater insight into the processes of the organism than that of the normal." … "Since disease process is a modification—and indeed, a very significant modification—of a normal process, biological research cannot afford to neglect it" [12].

4.4 Unawareness of lost function

In my doctoral dissertation, I described a patient with unilateral neglect of the *left side of the visual space*. This disorder is manifested by inattention or neglect of objects located on one side of a vertical midline. This type of neglect can take place in patients that have normal or near normal vision. The neglect can be easily tested when it involves the left side of the visual space by asking the patient to divide a horizontal line into two equal segments. When neglect is present, the patients divide the line, but leave a much longer segment on the neglected side. Another test consists in asking the patient to copy or draw symmetrical figures such as a flower or a clock. The distortion of their drawings is evident not only when copying a model (Fig. 4.5a), but also when the same object is drawn from memory (Fig. 4.5b).

A similar defect can be observed when patients draw a clock, as shown in Fig. 4.6. When I indicated that the clock was incomplete, the patient was unable to perceive that defect.

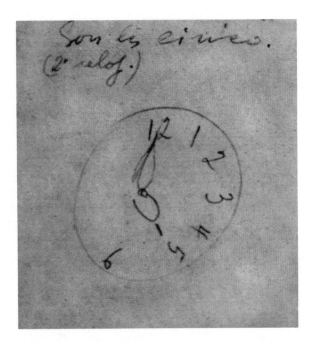

Fig. 4.6 A clock drawn from memory. Drawing of a clock by the same patient suffering from sensory neglect of his left visual field. As he indicated, "It's five o'clock" (in Spanish, "Son las cinco", second clock). The patient could not perceive that numbers 7–11 were missing from the clock

On my insistence, the patient added the number six, but was unable to see that numbers 7–11 were missing.

The devastation produced by bilateral brain lesions make it unlikely that patients could communicate their feelings. However, vascular or traumatic bilateral lesions to the visual areas (located in the posterior part of the brain) can spare language and produce blindness that may not be perceived by the patient. To understand this visual unawareness, we can compare the differences between spots of blindness produced by a lesion in the retina with that of the *physiological blind spot*, which is subjectively imperceptible. A lesion in the retina (also in the optic nerve or in the visual pathways) produces a distinct area of blindness, corresponding to its brain representation, surrounded by normal vision. Thus, the patient perceives clearly that there is a blind area. In contrast, we do not spontaneously perceive the physiological blind spot because this part of the retina *does not have a representation* in the brain that could register the lack of incoming impulses. In addition, vision by the other eye helps to cover up the lack of vision of the blind spot. We cannot miss a sensation for which we do not have a receptive area in the brain. This is why we do not feel that the back of our head is blind. Similarly, we do not miss the echolocation of a bat or the magnetic sense of orientation of some birds. Paraphrasing Thomas Nagel, we cannot even imagine what it is like to have the senses that we never had [13]. Thus, if the brain area that "represents" a body part is destroyed, the perception of the corresponding body part is lost together with the awareness of the loss.

The unawareness of brain diseases (anosognosia) is common in patients with arterial lesions of the visual areas of the brain. Bilateral lesions of the vision areas of the occipital lobes can produce central blindness that the patient does not perceive (Anton syndrome). The patients not only do not complain, but also deny their blindness, or attribute it to the lack of light or poor glasses. The lesion is sometimes associated with loss of visual memories, such that the patient cannot recall any visual image [6]. The unawareness of the blindness and the lack of concern about the disease have been attributed to some psychogenic "denial" syndrome. However, this is not the case, because most patients with lesions outside the brain recognize their diseases. The destruction or the temporary dysfunction of the higher visual centers is comparable to the natural absence of the specific sense. Thus, when a lesion involves the highest centers for a specific function, the function disappears without trace, and the patient remains unaware of the loss. Some patients cannot even think that they are blind, in the same way that we do not miss the bat's echolocation which our species has never possessed. However, the severity of the unawareness may fluctuate with the state of the lesion and with the surrounding inflammatory process.

Disorders of the body image are not so rare, if one looks for them. They are intriguing and provide an enormous wealth of precious information regarding brain function and human nature, and the control of willed or unwilled acts. Unfortunately, to obtain interesting information requires tedious and time-consuming tests that are not immediately beneficial to the patients. Modern brain imaging techniques are much more precise and expedient in localizing brain lesions, determining their nature, and suggesting the best course of action.

4.5 Split-brain patients have a split mind

The most astonishing examples of the fragmentation of the mind are those described in the split-brain studies. Roger W. Sperry and collaborators performed ingenious experiments in monkeys [14]. They cut the nerve fibers connecting the right and left-brain hemispheres as well as the crossed fibers of the optic nerves that allow one eye to send images to the other side of the brain. These animals could be trained with one of their eyes covered to perform complex visual tasks, but when tested using the other eye (that only projects to the other half of the brain) they failed the task. By contrast, monkeys in which the fibers connecting both sides of the brain were left intact can learn with one eye and respond successfully when tested with either eye.

Similar studies were performed in humans suffering from intractable epilepsy, who underwent surgery to stop the spreading of the seizures to both sides of the brain. The surgical procedure cut most of the fibers (corpus callosum and anterior commissure) connecting the right and left-brain hemispheres. Following the surgery, images rapidly flashed on the right or left visual fields could be perceived only by the right- or left-brain unless the patient was given time to move the eyes or the head and send the information to both sides of the brain. These experiments were designed to show that both sides of the brain could independently perceive, learn, remember, and respond. Thus, neither the split-brain patients nor the operated monkeys can transfer information from one side of the brain to the other.

As indicated previously, verbal language is localized in the dominant side of the brain, which in right-handed people is the left side of the brain. This explains why, in split-brain patients, the left-brain hemisphere "retains the capacity to speak its mind" whereas the right brain can neither speak nor write [14]. Surprisingly, the comprehension of spoken instructions is generally good by the right or subordinate hemisphere in right-handed persons. In addition, the right side of the brain outperforms the left in some spatial tasks and in nonverbal reasoning. In practice, these defects are not readily apparent in split-brain patients because normal visual clues (as opposed to rapid flashes on one visual field) are perceived by both sides of the brain. In these patients, the most appropriate part of the brain will respond faster, so the defect is almost imperceptible. However, some complex tasks cannot be performed after separating the two brain hemispheres, indicating that these tasks require the active cooperation of both sides of the brain. Thus, the concept of complementarities of both hemispheres has replaced that of one-sided dominance [14].

The most intriguing result of the split-brain studies is in showing that both sides of the brain are independently conscious when disconnected. Both hemispheres retain the highest form of consciousness: self-consciousness, but they behave as different individuals. However, all the intellectual and emotional responses to relatives, friends, and political and religious figures are maintained by both sides of the brain, indicating that these experiences were stored independently before the surgery. Moreover, both sides are aware of the past, present, and future. Thus, both sides of the brain have *independently conscious minds*, but after the surgery, they cannot fully communicate with each other. These results indicate that the mind is not a uniform spiritual substance, because it can be divided by the scalpel. The results of the operation also question the problem of personal identity and the unity of consciousness, because each hemisphere has a mind of its own. Actually, the evidence seems to indicate that *consciousness is modular*, but that under normal conditions, the apparent unity is maintained by the neural interconnectivity between all of its parts [14]. The division of consciousness is a fascinating observation that negates the dogma of the unity and indestructibility of the human soul.

4.6 Alzheimer's disease destroys the self

Memory is one of the most fundamental properties of the nervous system, without which the individual self, culture, and history simply would not exist. Forgetfulness of recent events is common to normal aging, but it is also one of the first symptoms of Alzheimer's disease, which may inexorably progress to destroy the most essential characteristics of a person. Alzheimer's disease has achieved recent notoriety because its frequency has increased as a result of people living longer. The public awareness of this devastating disease was also greatly increased when former president Ronald Reagan dramatically announced in November 1994 that he was in the early stages of the devastating disease. The onset is insidious, but the course is relentless. The disease produces deep mental changes that lead to the complete deterioration of the mind. As Glenn Collins put it in the New York Times in 1994, "the disease steals the soul" [15].

The brain lesions in Alzheimer's dementia consist of senile plaques and neurofibrillary tangles that result in neuronal loss and atrophy, predominantly in the brain's cortical association areas and in the hippocampal formation (see Fig. 7.3), located in the temporal lobe. Lesions in the temporal lobe explain the memory and cognitive deficits whereas the parietal lobe lesions explain the deficiencies in the visuospatial functions that are also affected in Alzheimer's patients. Modern neuroimaging techniques provide an indication of the brain atrophy, which is roughly correlated with the degree and the progression of the dementia. Transgenic mouse models of the dementia also are being increasingly used for development of new medications that may be effective in humans.

The initial symptoms of Alzheimer's disease include deficits in the short-term semantic (declarative or explicit) memory and in attention mechanisms. The progression of the dementia provides an additional demonstration of the modular organization of the human mind. When the dementia advances, patients not only forget names, but they also lose their abstract thinking and judgment, and cannot follow simple instructions, get dressed, or find their way in their own homes. Alzheimer disease patients become depressed or agitated, with emotional outbursts, insomnia, and language disorders. These problems are a constant source of embarrassment and irritation, not only for the patient, but also for the family that watches the inexorable disintegration of a loved one in despair. The constellation of symptoms varies initially in different patients, but progressively they all look the same. Patients fail to recognize their spouses and children, and become grossly self-centered. As the severity of the dementia increases, patients even forget their own names, cannot maintain their personal hygiene, and become incoherent. At some point, the victims of Alzheimer's disease cease to be human beings, despite their resemblance with their earlier physical images. They become living cadavers which, in losing their brains, also lose their self. Alzheimer's disease is the most dramatic and cruel demonstration that the mind is not a spiritual, indestructible, supernatural substance, but an aggregate of physical brain functions.

4.7 Reflections on nature's most amazing window

The observation of neurological patients with brain diseases provides the most amazing window into human nature. Paradoxically, we become aware of the complexity of the brain when drugs or diseases alter its normal functions. Brain diseases let us look at and analyze the mechanisms of mental processes. Normal functions seem deceptively simple, and we take it for granted when we are able to remember a word, identify an object by touch, or make abstractions in diverse situations. The brain is no doubt the *most complex* single object in the accessible Universe. To understand it, we take it apart—figuratively—and then put it together in our minds, as physicists do with the atom and the subatomic particles. We make hypotheses and test them, like all scientists that deal with complex objects.

There are obvious ethical limitations when dealing with human diseases. Brain surgery generally produces some collateral damage, as when surgeons split the brain to treat refractory forms of epilepsy, or resect brain areas invaded by tumors. Unfortunately, diseases have replicated the cruelest and most unethical experiments that one could imagine. Neurology keeps offering opportunities to peek through the most interesting window in the Universe. Neurologists have been the pioneers in studying human nature by looking at the effects of diseases on the brain—experiments made by nature—as Kurt Goldstein said.[4] A complementary approach has been the use of experimental animals, which have much simpler brains. However, the most characteristic human attributes are inaccessible through animal experiments.

Hughlings Jackson's studies provided some of the first indications that sensations, perceptions, emotions, and thoughts are not just "psychological" processes, but physical events that take place in the brain. Curiously, the elemental neuronal mechanisms that support different functions are of identical nature. He observed, "…the psychological, like the physical processes of the nervous system, can only be functions of complex combinations of motor and sensory nerves."

---

[4] An interesting book on cognitive neuroscience has been based on descriptions of patients and diseases [16].

Schoolchildren are taught that we have a body composed of several organs, one of which is the brain. It is a revelation to slowly learn that this notion is incorrect. We do not *have* a brain. We *are* our brain.[5] Actually, we are not the brain itself, but *a collection of functions* of our brain. In the neurology ward, it is evident that what happens to brain functions is the only true reality. "We" are aware only of what happens in "our" brains. The rest of our body is *external to our minds*, as if it were part of the external world. The mental image of our body seems natural and innate, but like the image of the external world, the body image is *created* by the active functions of the brain with the information provided by the senses.

The self is a collection of *modular* functions that must be integrated in time and space to achieve the unity that we take for granted. The perception of ourselves is composed of the multiple perceptions of our bodies and external world, the memories of experiences, our instincts and appetites, and our uniquely human feelings and aspirations for the future. The awareness of the multiple components of the self is not simultaneous; they occur at different times, clearly indicating that we are not static objects, but rapidly changing processes. Some of the functions that constitute the self are discretely localized, but most are distributed throughout the brain.

The disturbances of the higher mental functions produced by brain diseases are astonishing because they reveal that what we call the mind is not a single indivisible unit, as also shown by the split-brain patients. Becoming aware of the complexity and modular structure of the brain leads to the realization that *the self—or the soul—cannot be conceived as a simple indivisible substance*, as described in the theological works of the great doctors of the faith. Brain diseases show that we are not a homogeneous soul, but *the multiple functions of our brain*. Moreover, this *collection of physical processes* keeps changing. If the brain functions stops, we stop *being*, as happens under deep general anesthesia, during which our own self becomes imperceptible. Under this condition, we are technically dead, because we would not survive if the life sustaining equipment were accidentally disconnected. Our soul cannot manifest itself during deep general anesthesia, and our brain activity is minimal. In contrast to normal sleep or sedation, the perception of elapsed time during deep general anesthesia is suspended, and when we wake up, we are surprised that the surgery has already been completed. Elapsed time cannot be perceived when our internal clocks are not ticking. Unfortunately, the functional character of the mind and the self are usually not recognized explicitly. In making abstractions, we tend to *hypostatize*, that is to view functions and processes as if they were substances. Taking functions for substances (i.e., believing that we have a

---

[5] There is an interesting article by Andy Clark on this subject [17].

supernatural soul) is one of the most serious obstacles to understanding the mind and the self.

References

1. Jackson JH. On Epilepsies and on the After-effects of Epileptic Discharges. In: Taylor J, editor. Selected writings of John Hughlings Jackson. 1st. ed. New York: Basic Books, Inc.; 1958:135–161.
2. Berkeley G. Philosophical Works. 2nd ed. London: Rowman and Littlefield; 1975.
3. Clarke E, Jacyna LS. Nineteenth-century Origins of Neuroscientific Concepts. Berkeley: University of California Press; 1987.
4. Temkin O. The Falling Sickness: A History of Epilepsy from the Greeks to the Beginnings of Modern Neurology. 2 ed. Baltimore: Johns Hopkins Press; 1971.
5. Snyder SH. Seeking god in the brain–efforts to localize higher brain functions. N Engl J Med 2008;358(1):6–7.
6. Critchley M. The Divine Banquet of the Brain. New York: Raven Press; 1979.
7. Korein J, Musacchio JM. LSD and focal cerebral lesions. Behavioral and EEG effects in patients with sensory defects. Neurology 1968;18:147–152.
8. Critchley M. The parietal lobes. London: Edward Arnold & Co.; 1953.
9. Ay H, Buonanno FS, Price BH, Le DA, Koroshetz WJ. Sensory alien hand syndrome: case report and review of the literature. Journal of Neurology, Neurosurgery & Psychiatry 1998;65(3):366–369.
10. Fisher CM. Alien hand phenomena: a review with the addition of six personal cases. Canadian Journal of Neurological Sciences 2000;27(3):192–203.
11. Bundick T, Jr., Spinella M. Subjective experience, involuntary movement, and posterior alien hand syndrome. Journal of Neurology, Neurosurgery & Psychiatry 2000;68 (1):83–85.
12. Goldstein K. The Organism. A holistic approach to biology. New York: American Book Company; 1939.
13. Nagel T. What Is It Like to Be a Bat? In: Block N, editor. Readings in Philosophy of Psychology. 1st. ed. Cambridge, MA: Harvard University Press; 1980:159–168.
14. Sperry RW. Consciousness, personal identity, and the divided brain. In: Benson DF, Zaidel E, editors. The dual brain. New York: The Guilford Press; 1985:11–26.
15. Collins G. Enduring a Disease that steals the Soul. The New York Times 1994 Oct 10.
16. Sacks O. Luria and "Romantic Science". In: Goldberg E, editor. Contemporary neuropsychology and the legacy of Luria. Institute for Research in Behavioral Neuroscience. xi, 287 pp.: 1990:181–194.
17. Clark A. I am John's Brain. Journal of Consciousness Studies 1995;2(2):144–148.

5 Why qualia and consciousness seem mysterious

Summary Qualitative experiences (qualia) and consciousness seem mysterious, but are easily understood when we realize that they are *neural processes* that provide *language-independent information* about external objects and about the state of the organism. Thus, colors, sounds, smells, and emotions can be named, but they are ineffable because they cannot be transmitted through explanations. Experiences are neural processes that compress large amounts of information into messages that are directly perceived and understood by association to other experiences, which provide their *aboutness*. Qualitative experiences are phylogenetically determined and allow humans and organisms without language to navigate in their environment, communicate, and satisfy their biological needs. Consciousness is an active process maintained by oscillating neural activity, which can focus on specific subjects or produce the simultaneous awareness of multiple experiences.

5.1 The puzzle of experiences and consciousness

Experiences provide the "what-it-is-like" or the non-verbal qualitative representations of the internal and external worlds, that we perceive through our senses, emotions and internal needs. Conscious experiences are among the most primitive forms of biological awareness that give us information about our surroundings and ourselves. This information is essential for the individual to survive long enough to reproduce. We can infer from our contact with domestic animals that our early ancestors also had experiences before language originated, probably millions of years ago. Experiences are biological processes that cannot be transferred as such through language.

Philosophers frequently ask, "How could the *physical* brain give rise to *conscious experience*?" This question implies that some do not believe that consciousness and experiences are physical processes that take place in the brain, and that provide essential information about the organism and its surroundings *in the absence* of language. The impression that experiences are non-physical processes may have been produced by the imperceptibility and transparency of the brain and its functions. The puzzle about the nature of phenomenal experiences and conscious processes has been revived during the last few years in several books and articles [1–3]. In a more recent book, Chalmers [4] catalogues a large portion of the literature about consciousness and experiences, but he does not explain how what he believes is a non-physical consciousness could interact with the physical body without violating the laws of thermodynamics.

To further complicate the issue, Chalmers' theory of *property dualism* proposes that consciousness can have both physical and phenomenal properties. He emphasizes that the phenomenal properties of consciousness, the "what-it-is-like" that characterize experiences—as viewed from the first person perspective—are irreducible to physical properties, because experiences are ontologically independent [2, 4]. This is questionable because non-physical phenomenal properties have never been proven to exist, much less to have any kind of causal efficacy. Chalmers insists that the hard problem of consciences is that of explaining conscious experiences.

In contrast to Chalmers' property dualism, neuroscientists believe that *conscious processes,* including all phenomenal properties, are realized by well-studied neurophysiologic mechanisms that detect their state of individual functions and the needs of the organism. There is nothing mysterious in hunger, pain, thirst, the need for love, or avoidance of predators. The physical nature of all these functions is well understood, because they have been functionally characterized and are modified by diseases and by the administration of drugs that act on specific brain systems. Chalmers refers to some neurophysiologic processes as neural "correlates" of consciousness. However, it is seriously misleading to call the neural mechanisms of consciousness "neural correlates", because when the mechanisms that maintain consciousness are suppressed, consciousness is also suppressed. This means that we are in the presence of an identity.

The contrast between the subjective, apparently spiritual nature of experiences and the physical nature of the body has preoccupied philosophers and biologists since René Descartes (1596–1650), who championed the idea of the *duality* of body (*res extensa*) and spirit (*res cogitans*) or thinking substance [5]. Many people also believe that consciousness is a manifestation of a hypothetical soul or spirit, as taught by most religions. However, this view has been discredited not only by the neurobiological and scientific perspectives, but also by common sense observations. Despite our feeling that experiences are transparent and seem non-physical (see Sect. 5.1.3), they consist in neural processes that are easily disrupted by physical agents such as diseases and trauma, as well as by the ingestion of psychoactive drugs or alcohol.

One of the confusing features of the puzzle is that the brain, which senses the body and the external environment, is imperceptible to itself. There are at least two reasons for this; one is that the brain does not have to be sensitive because it is well protected by the hardness of the skull and by the sensitivity provided by the hair and scalp. The second reason is that the brain, as the last member of the sensory chain, must be itself insensitive to avoid the infinite regress implied in sensing the sensors that sense the brain sensors and so on (see Chap. 4). The first-person impression is that all brain functions are mysteriously realized by a supernatural soul or spirit, as discussed in Chap. 2. Despite these naïve feelings, all sensations, perceptions, and emotions are known to be carried out by the brain, even if the brain cannot sense itself. Neurology and neuroscience clearly show that qualitative experiences are neural processes by which we sense the properties of external objects and the internal state of the organism.

Qualitative experiences evolved millions of years before language, so they are language-independent and their message is directly perceptible as experiences such as pains, hunger, or emotions. They are hard to explain, not because they are mysterious, but because language was recently acquired and experiences actually give *meaning* to language by serving to ground words. Ineffable words get their meaning from the experiences that they name. Thus, these words can neither be explained nor understood by persons who have never had such experiences. This is why we cannot explain an orgasm to a very young child, or "what-it-is-like" to see red to a color-blind person by mentioning ripe tomatoes, stop signs, or sunsets. In essence, the words that name experiences cannot be understood by persons who have never had such experience. There is no doubt that there is a fundamental difference between the phenomenal view and the propositional explanation of consciousness and experiences [6].

5.1.1 The ineffability of experiences

The impossibility of explaining qualitative experiences is a vexing problem that has preoccupied philosophers for decades. As Dennett indicates, qualia are the paradigm of ineffable items [7]. This means that we cannot explain qualitative experiences, e.g., the what-it-is-like to see red to a color-blind person. Consider a classic example the imaginary case of Mary, a scientist that grew up in a closed room without ever seeing anything red. To compensate, Mary read many scientific papers that explained what happens in the brain when we see something red, but she still could not figure out what-it-is-like to see something red. Eventually, Mary was let out of the room and had the experience of seeing something red, which neither language nor science could explain clearly enough to *transfer* a novel experience. Some philosophers use this example to show that we are not intelligent enough to understand experiences [1], whereas others believe that experiences are facts that are not describable by words. As

Nagel said, to know what it is like to be a bat and navigate the world through echolocation, you must be a bat [8].

According to David Chalmers: "We have no independent language for describing phenomenal qualities. ... [T]here is something ineffable about them.... In talking about phenomenal qualities, we have to specify the qualities in question in terms of associated external properties, or in terms of associated causal roles [2]." He coined the term *hard problem* when he wrote: "The really hard problem of consciousness is the problem of experience." "How can we explain why there is something it is like to entertain a mental image, or to experience an emotion?" Chalmers wonders why physical processes give rise to a rich inner life. He believes that experiences cannot be reduced to neural processes, so he supports a property dualism, which postulates irreducibly mentalistic properties that cannot be produced by physical processes.[1] However, the real problem is quite different: it consists in the fact that verbal

Table 5.1 The many connotations of "Red"

Because of the transparency of phenomenal processes and the ambiguity of our language, several distinctions are essential for referring to the different meanings in which "red" can be used

***Objective*-red** is the objective light reflectance of ripe tomatoes, a non-internalizable physical process which consists in an electromagnetic radiation of wavelengths around 600–650 nm, and which is an intrinsic feature of the *external object*

***Neural*-red** is the internalized signal of the red light, which becomes an experience when it is incorporated into conscious processes. It consists in the neurophysiological activity elaborated in the brain by sensing red light. It is comparable to the *sensor data* of robotic systems and *AI*. *Neural*-red is what Mary presumably studied in her room. However, *neural*-red cannot be produced by propositional explanations

***Phenomenal*-red** is the sense-datum produced when the activity of *neural*-red is incorporated into conscious processes through binding. It is perceived and remembered as the what-it-is-like to see red; it is what Mary learned outside her black and white room. The experience can be described physiologically, but it cannot be duplicated in other brains through explanations. *Phenomenal*-red is both, a neural process and an ineffable qualitative experience that serves to anchor "red" in the verbal-phenomenal lexicon (Musacchio [6])

Phenomenal concept of seeing red is produced by experiencing what-it-is-like to see *objective*-red. This anchors the meaning of seeing red

The *aboutness* of experiencing red is a relational phenomenal concept that is object-dependent and is established through additional experiences, like seeing red fruit, red faces, red signs, etc., in which "red" has different meanings

The word red is anchored in the verbal-phenomenal lexicon by giving a name to the experience, phenomenal concept, and aboutness of seeing red (Chap. 6)

---

[1] David Chalmers [2, 9].

explanations of experiences cannot generate the physical processes that realize experiences in the brain of the listener. The physical realization of experiences is part of our phylogenetic endowment, which is essential to ground words and to develop a language. Moreover, the grounding of words in the common phylogenetic endowment is essential to translate languages.

Many philosophical problems have their origin in the lack of precision of our language [10]. For example, some verbs, such as to know, see, and feel, have different meanings depending on their context. A similar ambiguity, which is not trivial, can be found in the meaning of the properties of objects, such as colors. For example, "red" may refer to an *objective* property of the surface of a tomato (*objective*-red) or to what somebody sees as *phenomenal*-red in a veridical perception, afterimage, illusion, or hallucination (see Table 5.1). Thus, *red* refers either to the *objective property* of the surface of a tomato or to the *experience* of seeing something red. The experience of seeing something red (phenomenal-red) is mediated by the *binding* or incorporation of *neural*-red into conscious processes. However, the *intrinsic features* (*or the details of the neurophysiologic processes*) that realize neural-red are never perceptible to the subject. I will refer to the different aspects and properties of experiences using "red" as an example (See Table 5.1). As explained, qualia are part of the phenomenal-propositional lexicon in which *we anchor the ineffable words of our language*, which are those that name qualitative experiences.

5.1.2 The transparency of experiences

Qualitative experiences are said to be transparent because they convey the feeling that we perceive the qualities of objects and the needs of the organism directly, "as they really are", and without any interference from our perceptive machinery and language. There seems to be no separation between the experience of seeing a ripe tomato and ourselves. Transparency is one of the essential qualities of experiences, which explains not only its role, but also some of the puzzling characteristics of experiences.[2] Harman notes that we are only aware of the features of the object (the redness of the tomato—*objective*-red), but not of the neural mechanisms of seeing red (*neural*-red—the neural mechanisms of the experience itself) [10]. Brian Loar also observed that we seem to be directly aware of the properties of objects rather than the properties of experience itself [20]. The transparency of experiences is biologically necessary, but as Levine indicated, the side effect is that transparency increases our "intuitive resistance to materialism" [21].

---

[2] I have reviewed the transparency of experience in Refs. [6, 11]. See also Refs. [2, 12–19].

Experiences have an *aboutness or intentional sense*, a meaning that is transparently evoked by association to other experiences. Clearly, hunger is about the desire to eat. Experiences and their meanings bring us directly in contact with our biological needs and with the external world. The aboutness of experiences is what motivates us to pursue or avoid them. Transparency also means that we are not directly aware of the neural mechanisms that implement the experience [10]. The imperceptibility of the neural realizers (such as *neural-red*) has clear biological advantages, but it creates the highly misleading illusion that experiences and the mind are nonphysical entities. A cognitive system that has a limited capacity of perceiving its own hardware is simpler and avoids the infinite regress implicit in sensing the sensors and the analyzers, and so on. To be useful, the neural mechanisms that realize experiences *must be imperceptible* to the subject. Metaphorically, the lens of the eye—like the lens of the camera—*must be transparent*; the camera and the lens must not appear in the pictures. The cognitive value of experiences and phenomenal concepts would be downgraded if all their neural mechanisms were also perceptible to the subject. Experiences are biologically useful *precisely* because through them, the external world and the needs of the organism are immediately perceptible without irrelevant details.

Our incapacity to perceive the neural mechanisms of the cognitive system contrasts sharply with the richness of everyday experiences. The objects of the external world seem real and clearly *physical* as opposed to the quality of the experience, which is often thought of as mental, psychological, or spiritual, and distinct from physical experiences. This reinforces the illusion that cognition and the mind are *nonphysical*. However, contrary to our common sense psychology, science indicates that the presumed immateriality of the mind is an illusion. Thus, the anti-physicalist intuitions of popular psychology are derived from the imperceptibility of the brain mechanisms that realize experiences and not from the character of the mind itself. Sensing and being aware of all our brain mechanisms would be detrimental to our perceptions and survival.

The only way to stop the potential regress of a sensory (or phenomenal) system is to have a limited ability to sense the sensors. Sensing all the components of the cognitive system would create not only an unnecessary biological handicap, but also infinite regress. To be efficient, the brain machinery must be *imperceptible to itself* and devoid of biologically irrelevant complexity. Unlike the brain, the sensory organs have sufficient pain sensitivity for elemental protection and survival. The uniqueness of the brain in not sensing itself provides protection against the *proliferation of superfluous structures* and the *regress* implied in sensing the sensors *ad infinitum*. The brain is insensitive, as indicated above, but it has an exquisite sensitivity for its essential nutrients, which are constantly monitored and controlled, such as the concentration of oxygen, glucose, and hormones in the blood.

The downside of simplicity and the price for the efficiency of our perceptive machinery is that, through introspection, *we cannot perceive the inner workings of the brain*. Thus, even if the imperceptibility of the basic neural processes provides a biological advantage, the resulting illusion is *a serious philosophical handicap* that drastically diminishes the value of introspection and phenomenology as exclusive methods for exploring our nature. This implies that the cognitive limitations of the subjective perspective must be supplemented by the empirical knowledge provided by science.

5.1.3 Conscious processes are maintained by specific activating systems

The contradictory opinions that originate from different philosophical or non-scientific approaches indicate that these approaches are not capable of defining experiences or conscious processes. In contrast, science and medicine have proven their effectiveness in producing consistent and reliable explanations that are verifiable by different observers. The biological bases of conscious processes have been established using a variety of different procedures and approaches, and we often find deep insights coming from unexpected places, as illustrated by the following story. The legendary neurosurgeon Wilder Penfield, in his efforts to remove brain areas that caused epileptic seizures without disturbing the speech mechanisms (Chap. 4), operated on patients using only local anesthesia so he could communicate with them during the operation [22]. Penfield probed the regions surrounding the epileptic focus with electrical stimulation while talking to the patients before removing any tissue and mapped the *sensory* and *motor* homunculi (Figs. 4.3, 4.4). He knew that the electrical stimulation of some of the brain areas that control language inhibits speech. In the process, Penfield also observed that when the base of the brain is disturbed, near the subthalamic nucleus (Fig. 5.1) patients might immediately lose consciousness. In contrast, he could remove large areas of the cerebral cortex in a conscious patient using only local anesthesia to open the skull. Curiously, no pain is perceived because, as we said above, the brain cannot sense itself, and the patients remain unaware of any change until asked to turn their attention to certain parts of the body, or to perform certain tasks that require the use of the removed brain region.

The understanding of the brain mechanisms that sustain attention and consciousness was also greatly advanced by previous studies showing that electrical stimulation of the brain stem in lightly anesthetized cats produces the electroencephalographic pattern of arousal that is characteristic of attention and alertness [23]. Lesions of the same regions in monkeys also produce various coma-like states that resemble deep sleep. Some of the neural networks that control arousal and attention are embedded in what was initially known as the *ascending reticular activating system*, shown in Fig. 5.1 [23]. In higher animals, including humans, this neuronal network reaches the intralaminar

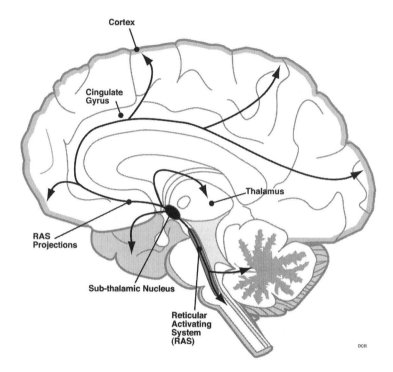

Fig. 5.1 Schematic drawing of the reticular activating system. The reticular activating system (RAS) consists of neuronal networks originating in brainstem regions that project upward to the subthalamic nucleus and from there to many cortical and subcortical brain structures as indicated by arrows. The RAS is responsible for maintaining conscious activity and is very sensitive to hypnotics and general anesthetic

nucleus of the thalamus, whose outputs can produce excitation in almost every brain region. These complex networks of neurons are involved in behavioral arousal, attention and sleep, as well as in the regulation of motor and autonomic reflexes [24–26]. These systems are similar across species and can be depressed or stimulated by the same drugs that are effective to induce sleep or alertness in humans.

Higher animals—including chimpanzees and domestic animals—also have night and day circadian cycles, as well as basic feelings and perceptions that are similar to ours. There is no doubt that these animals have some understanding of their own world or that, at least in a limited sense, they can communicate with us and with each other using signals that they learn to interpret. However, animals—perhaps with the exception of whales—do not have the complex language that could help them understand each other and improve their knowledge of the environment. Thus, how do they manage to satisfy their basic needs and communicate? The answer is probably through observations and experiences. As discussed in the next sections, qualitative experiences are *functional equivalents* or neural models for objects and processes that *cannot be internalized* by the brain or that represent internal states of the organism, such as hunger or pain.

5.2 Sensing and understanding internalized information

During childhood and adolescence, almost everyone believes that they perceive the world as it is. However, experience later reveals that neither the objects nor their properties are directly perceived as they are. Similarly, it was assumed by early philosophers and scientists that we could internalize *images or impressions* that faithfully represented the external world. The problem with this early assertion is that it needs a mind's eye or an internal observer that must also internalize images, a situation that implies an infinite chain of observers and regress.

Despite their limitations, qualitative experiences are the most basic and essential forms of acquiring information, without which we would not be able to have any knowledge. In the early stages of our cultural development, humans would have operated in an automatic fashion, *without really under-standing what was being sensed*. However, knowledge of the surrounding space, as well as the perception of time and movements improved rapidly because of better knowledge of the environment. The cultural transmission of knowledge in higher animals and humans was crucial for competitive survival and natural selection.

The illusory nature of some sensations and perceptions was already known to early Greek philosophers. As Aristotle (384–322 BC) already thought, we can perceive three-dimensional space and temporal intervals through more than one sensory modality, such as vision, touch, and hearing. However, most other perceptions, such as taste, smell, and color depend on where in the brain the sensory stimulation arrives. Thus, the what-it-is-like of the experience is provided by the region of the brain that receives the stimuli. Even though sensing is essential for knowledge, not all philosophers seem to be concerned about how we physically internalize information through the senses. To under-stand sensing, we cannot rely only on introspection. First, we must be aware of the biology of our senses and cognition. Language is not necessary for under-standing the meaning of most experiences, because experiences are unique and refer to other experiences. Together, these associated experiences form a cogni-tive network that is limited, but essential and sufficient to sustain conscious life and to promote survival in a language-independent fashion. The ability to communicate without language is also quite evident in babies and small infants. Qualitative experiences are *language independent*, but as discussed in Chap. 6, experiences are essential prerequisites to anchor words and to develop a verbal language.

5.2.1 Sensing the external world

Sensing the external world and sensing our bodies are important cognitive abilities, which developed before anyone knew about knowledge, philosophy, or language. Today, as evolved creatures with a language, we can reflect and say with confidence that animal survival has been made possible by sensing the environment and by knowing about food and predators. In contrast, the survival of plants, bacteria, and lower animals is based on their enormous reproductive ability.

There are some facts that are not evident to the subjective perspective and that should be taken into account to understand sensing [27]. Our brain can only internalize nerve action potentials (Fig. 5.2) produced by specialized sense receptors.

Thus we cannot experience the external world directly. In contrast to telephones and televisions, the information elaborated by the senses and entering the brain cannot be transformed back into the original stimuli (such as voices or light). Thus we must understand the world indirectly, through the neural signals internalized by the different senses. Understanding is achieved only through several additional processes, such as getting information through other senses and taking into account the reference (aboutness) provided by associated experiences.

The problem of acquiring information through the senses without a language was solved by Nature with the development of qualitative experiences that were initially used to navigate the environment and to deal with prey and predators. We know today that sensations or qualitative experiences are an *absolute necessity* for animal survival; they are also essential for developing other forms of knowledge and the capacity to reason using experiences, images, words, and symbols. However, many—and perhaps all—qualitative experiences *falsify* reality, because we actually do not perceive things as they are. The *what-it-is-like* that we experience is not identical to the *what-it-is*. We manage to survive only by *learning to interpret* the meaning of our experiences. However, before getting into the aboutness of experiences, I will discuss the problems of sensing and internalizing information.

Despite the many difficulties in understanding sensing, we know that knowledge is possible, even though limited and distorted. The empirical success of being able to modify our surroundings gives us additional confidence that the world is cognoscible, even if our knowledge is initially indirect and we cannot perceive reality as it is. The crucial question is whether *all experiences* provide direct and genuine knowledge of the properties of external objects and organic needs, as a realist would think, or if it only *seems* that they do. The evidence is contradictory: on one hand, we know that we can accurately perceive shapes and the space that immediately surrounds us. On the other hand, we know that we are subject to illusions, although we tend to believe at

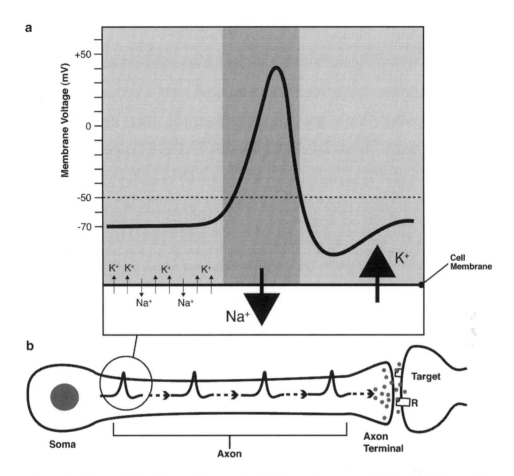

Fig. 5.2 Illustration of an action potential. An action potential (**a**), which is part of a train of action potentials along the axon of a nerve cell illustrated in (**b**). (**a**) Shows that there is a constant flow of Na+and K+through the nerve membrane. When a nerve impulse takes place (upward curves), a train of action potentials can be measured by registering the electrical activity inside the nerve (axon) with a capillary pipette, which acts as a recording electrode (not shown). The action potential is produced by the rapid entry of Na+to the nerve (large arrow down), which is followed by a compensatory, slower exit of intracellular K +(large arrow up). (**b**) Shows the propagation of the nerve impulse from the cell body (soma) through the axon in the direction of the axon terminal. The arrival of the action potential at the nerve ending results in the release of neurotransmitter molecules (small grey circles) that bind to receptors (R) located on the surface of the target cell. The target cell is stimulated or inhibited, depending on the neurotransmitter and receptors involved

first impression that things are just the way they seem to us. For example, most people do not know that colors are made up by the eyes and the brain and are not the exclusive properties of the external objects. Thus, qualitative experiences are internal processes that are not identical to what exists in the external world. However, some of our more objective experiences, such as the perception of time and space are somehow expressible by referring them to minutes or hours or to inches and yards. For example, we measure time and space by standards that are external to us, whereas hot or salty are ineffable terms because they are grounded on qualitative experiences that cannot be rigorously quantitated. We have all experienced that some foods or drugs change the taste of the next food we eat. The fact that experiences are modifiable by diseases, surgery, or drugs indicates that they are physical processes taking place in our brains.

5.2.2 Sensing the internal world

In contrast to the qualitative experiences generated from the external world—*exoqualia*—we have direct access to the endogenously generated qualia—*endoqualia*—that are unique experiences, such as thirst, hunger, satiety, anxiety, and sexual desire. Endoqualia are generated by the needs of the organism and by its relation with the external environment. They are contingent cognitive shortcuts that lack intrinsic meaning, but have high survival value, because they are associated to essential organic needs. Some experiences, however, such as emotions and pain, are innately hardwired, so they can rapidly trigger biologically advantageous responses. In lower animals, some of these states might not even be conscious. The physical agents that trigger *exoqualia* are easily identifiable in general, but the task is not so simple for *endoqualia*. Such is the case of experiences that refer to organic needs (hunger, thirst) or to the self, such as fear, love, depressive feelings, and other emotional states. These experiences are highly influenced by innate components, such as the sexual drive. All these considerations indicate that internally generated qualitative experiences are a direct, *nonverbal, phenomenal language* by which all organisms are motivated to satisfy their organic needs, *even without knowing what they are doing*. There is no doubt that endoqualia—all our desires—are major factors that influence most forms of human and animal behavior and shape our culture and advertising industry.

Today, we know that emotions are physical states that process information about the organism in its relation to the environment. Unknown through the subjective perspective, all emotional processes are realized in specific neural circuits, which are mediated by changes in neurotransmitters, hormones, and other cellular messengers. The effects of drugs, electrical brain stimulation, and anatomical lesions, as well as electrophysiological and fMRI studies clearly indicate that emotions are physical processes taking place in the brain [28]. These aspects as well as the neurobiological bases of fear have been thoroughly examined in several reviews and recent popular books by Joseph E. LeDoux,[3] that clearly implicate the amygdala, in the anterior part of the temporal lobe, as the main site where fear and other emotions are processed (see Fig. 7.3).

5.3 Experiences require binding for integration into consciousness

Our intuitions cannot be used as reliable sources for understanding either the nature of experiences or how the brain operates. All intuitions are potentially misleading, so the problem of the nature of experiences must be approached from a scientific perspective. Besides, we need a vocabulary that implies neither images nor internal observers.[4] The internalized information contained in neural processes that form a core of the sensation must be incorporated into conscious processes to produce an experience. We sense the internalized information as colors, sounds, or smells, but we cannot sense either the nerves or the brain that provide the intrinsic features of the experience. If the neural processes necessary to realize experiences were not bound into conscious experiences, these experiences would be considered "unfelt pains", unattended sensations, unconscious processes, etc. Stimuli that remain unconscious, even for a short time, would account for the automatic responses to sensory input that take place before we could perceive them. In addition, the physicality of all neural and mental processes dissolves the problem of the "mental" causation, which consists in explaining how the "mental" (in the non-physical sense) and the physical could interact.[5] We do not have a detailed answer to how neural encoding takes place, but there is a lot of information on the processes involved [38–42].

Through the subjective perspective, we have the undeniable impression that we can hold images in mind, dream about them, and recall them at will, but as indicated, this requires internal observers. Thus, the question that remains unanswered is: How is it possible to experience "mental images" when there are no mental images in the brain? There seems to be no doubt that the information is stored mainly in the cerebral cortex (Fig. 4.2) and subcortical circuits of the brain, but there are no indications on how the information is incorporated into conscious processes. Graphic images seem to require an individual observer and create some kind of Cartesian theater and regress.

---

[3] Joseph E. LeDoux [29–32].

[4] The anthropomorphism of our terminology is so ingrained in our language that it is difficult to find common words that do not imply any regress. The best option is to use technical descriptions of the neurobiological processes that have been obtained through the third person perspective. However, the task is not easy because words that refer to phenomenal experiences are ineffable [33], so they only can be described or understood from the neurobiological perspective.

[5] The problem of mental causation is discussed in Refs. [18, 34–37].

However, this is not necessary, since in normal perceptions, the neural models produced by sensations seem to be directly incorporated into conscious processes. The incorporation into conscious processes takes place by synchronization of the electrical oscillations at "40 Hz.", or in the gamma band of 30–70 Hz. The idea that perceptions may require an internal observer is a carryover from Descartes' assumption about the existence of internal images [5].

As indicated previously, neurologists and neurosurgeons discovered that small tumors or cysts in the base of the brain or in the pituitary gland produced progressive loss of consciousness and coma, which in some cases were reversible after an operation. Additional studies made it clear that the mechanisms of wakefulness, sleep, and maintenance of consciousness take place through synchronization of an activating system that includes multiple brain regions.[6] We now know that wakefulness is associated with a low-amplitude, high-frequency electroencephalogram, whereas deep sleep (physiological unconsciousness) is characterized by high-amplitude, low-frequency waves [43, 45, 49]. In addition, the level of consciousness during anesthesia can be accurately predicted by sophisticated analysis of the electroencephalogram [50].

Many of the recent studies on cognition have been conducted on the visual system, probably because this system has the greatest capacity to internalize high volumes of information in an almost isomorphic fashion. Processing of visual information starts in the retina, where specialized cells capture different elements of the sensation and project them to the brain's occipital cortex through two major pathways[7] (Fig. 5.3).

From the occipital cortex (Fig. 5.3a), parallel pathways convey information to the posterior parietal cortex (dorsal pathway), which provides information about the location of the object, and to the inferior temporal areas (ventral pathway), which provides information about the identity of the object. This is an oversimplification, but what is important is that the information is not projected on the brain like a photograph. Color, motion, depth, shape, contours, distance, etc. are processed in multiple cortical areas in a parallel distributed processing. Therefore, the neural models corresponding to the different elements of the retinal image must be brought together again by *multiple stages of binding* [52]. To complicate things further, color and form are processed almost simultaneously, but movement perception is delayed about 50 ms [53]. This implies that the subjective coherence of the visual perception is dependent on several analytical processes, the results of which must be reconstituted in time and space. Again, as stated by Kandel and Wurtz [26],

---

[6] Additional information and references can be found in Refs. [11, 23, 25, 43–48].

[7] The complexity of the visual system is enormous and cannot be summarized in the space available. For additional information, the reader should consult [51] or [26].

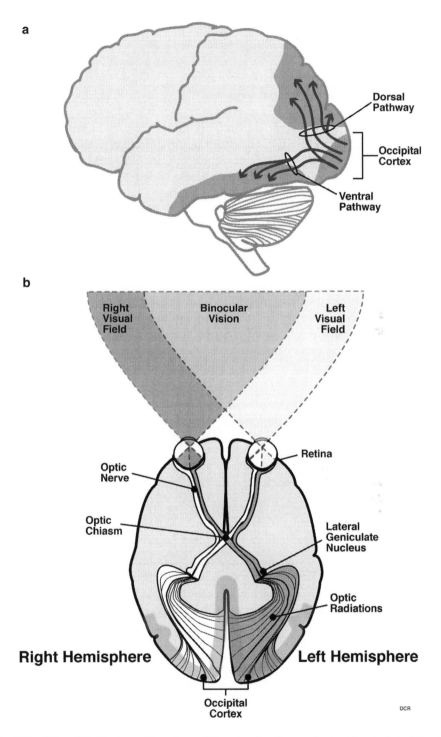

Fig. 5.3   Visual Pathways. Drawings of the visual pathways by looking in (**a**) at the left side of the brain, and in (**b**) by looking from below at the ventral surface of the brain. In (**a**), the visually activated portions of the occipital cortex are shown to project through the dorsal

visual perception is a *creative* process. The same could be said about the binding that creates the self, the dynamic process that constitutes us as we are.

5.4 How qualia acquire their meaning

Qualitative experiences acquire their meaning by association to other experiences that provide their *aboutness* or reference; this allows organisms without language to navigate the environment and to satisfy their biological needs. Infants quickly learn what to do when they feel thirsty, hungry, or in pain. Actually, there is evidence that the aboutness of most experiences may be innate, especially in some animal species that are born more mature than humans are and seem to know what to do immediately after birth. Even so, all animals including humans learn progressively more about how to interpret experiences and what to do about them. This knowledge is language indepen-dent, so their understanding is what gives origin to *phenomenal concepts* (see Chap. 6). Thus, most endoqualia are spontaneously associated to other experiences. For example, thirst produces the desire to drink water, even if the subject does not understand the nature of dehydration. Animals learn empiri-cally that an uncomfortable feeling can be eliminated by certain actions. The empirical association between an experience and its aboutness is obviously sufficient for survival under natural conditions, even if it is not enough for true understanding.

We can distinguish several kinds of aboutness; for example, *innate aboutness* and *phenomenal aboutness*, which are common to all animals. The *innate aboutness* of some experiences is hard-wired and is characteristic of some instinctual reactions, which may not necessarily be conscious. Encountering a fear object such as a snake is an example of a primary inducer of fear [54]. Rats and humans are both frightened by the sight of snakes. Similarly, infants and newly born mammals are hard-wired to suckle, and some baby birds have an instinctive reaction to hide or escape at the view of predators [55]. Laboratory-reared rats will also either freeze or try to escape in the presence of a cat, even if

---

Fig. 5.3 (continued) and ventral pathways. The dorsal pathway is mainly concerned with the location of objects within the visual field, whereas the ventral pathway projects toward the ventral temporal region and is mostly concerned with the recognition of faces and objects. In (**b**), the view of the brain from below shows that the central portion of the visual field is projected to both sides of the brain, providing right and left independent images of the same object, which generates binocular or stereoscopic vision. In contrast, the extreme right and left visual fields, which provide peripheral vision, project onto the nasal side of the same side (ipsilateral) retina, whose fibers pass through the optic nerve and cross the midline in the optic chiasm, before reaching the lateral geniculate nucleus, optic radiations, and occipital cortex of the contralateral brain hemisphere

they have never seen one before [31, 56]. These are all prime examples of innate aboutness or hard wired instinctual reactions.

Phenomenal aboutness is established through additional experiences and its value resides in having a *language-independent capacity to refer* (Table 5.2). Phenomenal aboutness is the most basic mechanism that animals have to make intelligent choices and to relate to each other. With the exception of a few experiences that have innate aboutness, most experiences do not have any predetermined semantic content until their meaning is learned by association with other experiences. This is clearly true for animals without language, which use phenomenal aboutness to navigate the world. One of the best-studied examples of learned phenomenal aboutness is that of fear conditioning [29, 31]. Phenomenal aboutness *corrects the lack of intrinsic meaning of most experiences*. The aboutness of experiences is learned implicitly by animals, as in classical and instrumental conditioning. The bell ringing before the meals meant for Pavlov's dogs that food was coming, so they started secreting gastric juice. For Skinner's rats, the view of the lever in the cage meant that they could get food by pressing the lever previously associated with food delivery. We all know that domestic animals readily show their understanding when we start to

Table 5.2   The fundamental differences between phenomenal and propositional knowledge (Modified from Musacchio [6])

Phenomenal knowledge	Propositional knowledge
• *Language-independent* neural models that are associated to their aboutness and serve to ground ineffable words	• *Language-dependent*, highly symbolic and translatable propositions. Language includes ineffable and partially explainable words
• Consists in *phenomenal* concepts (the what-it-is-like) generated by qualitative experiences	• Consists in *propositional* concepts acquired through language and layers of symbolism
• Implemented by mostly *hard-wired*, innate circuits, which like the visual system, require usage to fully develop their potential	• Innate capacity of *highly plastic* circuits, that use language and symbols developed through cultural influences and practice
• Self-sufficient	• Phenomenal knowledge-dependent
• *Ineffable*, private. Refers through the associated aboutness	• *Explainable* through several levels, public. Refers through propositions
• Common to higher animals and humans	• Exclusively human
• Implemented in phylogenetically old, anatomically well-determined structures, which are symmetrically distributed on both sides of the brain	• Implemented mainly on one side of the brain, in the dominant hemisphere, around the perisylvian fissure and in other phylogenetically recent structures
• *Concrete thinking* and language-independent analogue reasoning	• *Abstract thinking* and highly symbolic reasoning that can utilize several languages

prepare their food. The meaning of experiences is an essential component of a virtual language-independent *phenomenal-intentional lexicon,* in which the "entries" are the what-it-is-like of experiences and their "definitions" consist in their aboutness and in their indexical reference [33] (see also Table 5.2).

The *indexical reference* consists in relating an observation (a sign of present or past occurrence) to another event or consequence that naturally follows [57]. Calls, gestures, attitudes, and facial expressions (laughing, crying, etc.) are generated in some organisms, probably without the initial intent to communicate. However, they can be interpreted as *predictors* of behavior. This indexical form of reference is more complex than learning the aboutness of the organism's own phenomenal states, such as thirst and hunger. Indexical reference is an interesting concept which has not been fully developed, and which has untapped heuristic value to understand the enormous power of non-verbal thinking. The observation of natural processes may have an indexical value that can be used to infer rudimentary forms of causality, logic, and elementary mathematics in the absence of a verbal language.

The what-it-is-like of experiences that we perceive immediately and the associated what-it-is-about, are both cognitive signals that allow for intelligent choices and integrated responses. This is true even if most experiences have no intrinsic meaning and falsify reality, or if we do not know exactly how they are realized. However, the integration of the neuroscience perspective into the philosophical discourse allows us to conclude that phenomenal concepts and their aboutness are instrumental in determining behavior. The capacity for choosing the right alternatives also fosters the development of higher intelligence, because the most intelligent (and devious) individuals are the most likely to succeed in leaving descendants. The aboutness of experiences also makes it possible to develop signals, language, and other symbolic forms that open the doors for higher forms of knowledge. *Propositional aboutness* is characteristically human and is established through verbal explanations.

The lack of association between the phenomenal character "the what-it-is-like" of experiences and their aboutness becomes evident in some emotional processes in which the agents that produce the experience cannot be identified subjectively, such as in anxiety, panic attacks, or endogenous depression. These abnormal processes are qualitative experiences without subjective aboutness, or "false alarms". In these cases, the independence of phenomenal character (sadness and depressed mood) and reference (what the depression is about) has been empirically corroborated by the success of antidepressant drug therapy and the absence of a reason to be depressed. This indicates that the feeling of sadness could occur as a brain process independently of any subjective reason.

All these considerations indicate that qualia, phenomenal concepts (the what-it-is-like), and the establishment of their reference are the most basic and essential elements to build knowledge. This serves to ground words in a virtual dictionary, the verbal-phenomenal lexicon, which is analogous to a bilingual dictionary that makes language and propositional knowledge possible (Table 5.2). Thus, the meaning of language is actually grounded on experiences (see Chap. 6 and Table 6.1 for discussion of the word-grounding problem).

5.5 How experiences generate the self

Magnetic tapes, DVDs, and memory chips encode images and sounds in a way that we cannot perceive without the appropriate playback machinery. In analogy with gadgets, we could reason that, in addition to the sensory mechanisms that encode information in the brain, we need a system to decode the information perceived or stored in order to recall, dream, or hallucinate. However, we must resist the temptation of postulating an "observing self", which would imply a mind's eye or a Cartesian theater and regress. Thus, whatever we internalize must be incorporated through binding into conscious processes, suggesting that there must be neural processes or models that encode each sensation, perception, or recollection.

The self could be conceived as the integration of the innate phylogenetic endowment of the species with the repository of our life experiences and aspirations. We are the functions of our brain, a collection of processes in constant change. Past memories are integrated with present experiences in an always-changing modular, dynamic entity. The ever-changing self derives its versatility from the binding of constantly changing brain processes. The self is composed of brain modular activities that are distributed and integrated or bound together in time and space by the general mechanisms of awareness, which include the activity of the brain systems that synchronize neuronal oscillations [25, 48]. The modularity of the self is evident from the multiple examples of personal fragmentation produced by neurological lesions (see Chap. 4). We are aware of our physical body, not because it belongs to us, but because it is wired and connected to "our" brain. The brain internalizes and integrates all information coming from our body, which becomes part of the body scheme [58, 59] and the self.

We are "our" functional brain, meaning that the self cannot be different from the functioning brain in which those processes take place. Thus, it is incorrect to say, "I have a brain". Instead, I should say, "I am the current state of my brain". One of the problems is that language reflects the bias of the first person perspective, which cannot perceive the physicality of the conglomerate of functions and states that we are. The self is perhaps the most complex function of the brain. Occasionally, we adopt a dissociative or schizoid way of referring to "our brain", as if it were not part of our own self or as if our own brain did not have anything to do with the genesis of our experiences or our thoughts. I can lose a limb, but not even a fraction of my brain and continue to be myself. My experiences and my thoughts are the physical processes that take place in the brain that I am. Actually, we are some of the ever-changing functions of "our" brains.

Caring for the self is encoded through aversive and reinforcing emotional experiences that result in the preservation of the integrity of the system through feedback mechanisms. This results in the survival of those organisms that are able to make the "best" choices. The unity of the self is dependent on the integrity of the brain. A variety of brain lesions and the consequences of cutting the corpus callosum, a fiber bundle connecting the right and left brain hemispheres and other brain connections in humans [59, 60], demonstrate the modularity of the self (see Chap. 4). Moreover, the self disappears entirely during deep surgical anesthesia, deep coma, and most likely after death.

All these observations have important philosophical consequences, because they imply that (ontologically) *consciousness and the self are neither a thing nor a substance*, but a collection of processes that include sensations, perceptions, and memories. In other words, consciousness and the self are a collection of *dynamic processes*, which incorporate not only the current experiences, but also all our current thoughts, memories, and emotional states. The dynamic character of the mechanisms of awareness is able to focus on current perceptions, phenomenal or propositional concepts, or on any aspect of the self, thus eliminating the need to postulate different *kinds* of consciousness. This is why *conscious processes* refer better than *consciousness* to what we are. An easy demonstration of the dynamic nature of the self is provided by the administration of a rapidly acting intravenous anesthetic, such as profolol or pentobarbital. These anesthetics produce a rapid loss of consciousness and changes in the electrical brain activity that is indicative of the level of anesthesia.

In summary, this chapter reviews the neural mechanisms underlying the apparent mystery of consciousness and experiences. The physical nature of these processes explains the impossibility of duplicating conscious processes in other people through explanations. A key concept is that qualitative experiences are neural processes that function as internal models of the external world and of the internal states of the organism. To perceive an experience, qualitative experiences must be integrated into conscious processes through the 40 Hz thalamo-cortical rhythmic activity (Fig. 5.1). A major conclusion is that consciousness and experiences are physical processes that take place in the brain. Previously, brain mechanisms were considered as non-physical or spiritual events, because they are imperceptible to the subject. Actually, the perception of the brain mechanism in all its complexibility would be detrimental to the perception of the external world and what is essential for survival (such as hunger, pain, or predators). Thus, the neuronal mechanisms that generate our feelings, sensations and perceptions are transparent to avoid infinite regress (see Chap. 4).

References

1. McGinn C. The Mysterious Flame: Conscious Minds in a Material World. First ed. New York: Basic Books; 1999.
2. Chalmers DJ. The Conscious Mind. 1st. ed. New York: Oxford University Press; 1996.
3. McGinn C. The Problem of Consciousness: Essays towards a Resolution. First ed. Cambridge, MA: Blackwell Publishers, Inc.; 1991.
4. Chalmers DJ. The Character of Consciousness. First ed. ed. New York: Oxford University Press; 2010.
5. Descartes R. Discourse on Method and The Meditations. London: Penguin Books; 1968.
6. Musacchio JM. Why Qualia and the Mind Seem Nonphysical? Synthese 2005;147: 425–460.
7. Dennett DC. Quining Qualia. In: Block N, Flanagan O, Guzeldere G, editors. The Nature of Consciousness. 1st. ed. Cambridge, MA: The MIT Press; 1997:619–642.
8. Nagel T. What Is It Like to Be a Bat? In: Block N, editor. Readings in Philosophy of Psychology. 1st. ed. Cambridge, MA: Harvard University Press; 1980:159–168.
9. Chalmers DJ. Facing up to the problem of consciousness. Journal of Consciousness Studies 1995;2(3):200–219.
10. Harman G. The Intrinsic Quality of Experience. In: Tomberling JE, editor. Philosophical Perspectives, 4. Action Theory and Philosophy of Mind, 1990. first ed. Atascadero, CA: Ridgeview Publishing Company; 1990:32–52.
11. Musacchio JM. Dissolving the Explanatory Gap: Neurobiological Differences Between Phenomenal and Propositional Knowledge. Brain & Mind 2002;3(3):331–365.
12. Block N, Stalnaker R. Conceptual analysis, Dualism and the Explanatory Gap. The Philosophical Review 1999;108(1):1–46.
13. Dennett DC. Consciousness Explained. 1st. ed. Boston, MA: Little, Brown and Company; 1991.
14. Jakab Z. Ineffability of qualia: A straightforward naturalistic explanation. Consciousness and Cognition 2000;9(3):329–351.
15. Martin MGF. The Transparency of Experience. Mind and Language 2002;17(4):376–425.
16. Metzinger T. The problem of Consciousness. In: Metzinger T, editor. Conscious Experience. First ed. Thorverton, UK: Imprint Academic/Sconingh; 1995:3–43.
17. Papineau D. The Antipathetic Fallacy and the Boundaries of Consciousness. In: Metzinger T, editor. Conscious Experience. 1st. ed. Thorverton, UK: Imprint Academic/Schoningh; 1995:259–270.
18. Papineau D. Thinking About Consciousness. New York: Oxford University Press; 2002.
19. Tye M. A Representational Theory of Pains and Their Phenomenal Character. In: Block N, Flanagan O, Guzeldere G, editors. The Nature of Consciousness. Cambridge, MA: The MIT Press; 1997:329–340.
20. Loar B. Phenomenal States. In: Block N, Flanagan O, Guzeldere G, editors. The Nature of Consciousness. 1st. ed. Cambridge, MA: The MIT Press; 1997:597–616.
21. Levine J. Materialism and Qualia: The Explanatory Gap. Pacific Philosophical Quarterly 1983;64:354–361.
22. Penfield W. The Second Career. Boston: Little, Brown and Co.; 1963.
23. Moruzzi G, Magoun HW. Brain stem reticular formation and activation of the EEG. Electroencephalogr Clin Neurophysiol 1949;1:455–473.
24. Joliot M, Ribary U, Llinas R. Human oscillatory brain activity near 40 Hz coexists with cognitive temporal binding. Proceedings of the National Academy of Sciences of the United States of America 1994;91(24):11748–11751.
25. Llinas R, Ribary U, Contreras D, Pedroarena C. The neuronal basis for consciousness. Philosophical Transactions of the Royal Society of London – Series B: Biological Sciences 1998;353(1377):1841–1849.

26. Kandel ER, Schwartz JH, Jessell TM. Principles of Neural Science. 4 ed. New York: McGraw-Hill; 2000.
27. Gardner EP, Martin JH. Coding of sensory information. In: Kandel ER, Schwartz JH, Jessell TM, editors. Principles of Neural Science. 4 ed. New York: McGraw-Hill; 2000:411–429.
28. Vrij A, Fisher R, Mann S, Leal S. Detecting deception by manipulating cognitive load. Trends in Cognitive Sciences 2006;10(4):141–142.
29. LeDoux JE. Emotion: Clues from the brain. Annu Rev Psychol 1995;46:209–235.
30. LeDoux JE. Emotion Circuits in the Brain. Annual Review of Neuroscience 2000;23 (1):155–184.
31. LeDoux JE. The Emotional Brain. 1st ed. New York: Simon and Schuster; 1996.
32. LeDoux JE. Synaptic Self. How our brains become who we are. First ed. New York: Viking; 2002.
33. Musacchio JM. The ineffability of qualia and the word-anchoring problem. Language Sciences 2005;27(4):403–435.
34. Velmans M. How Could Conscious Experiences Affect Brains? Journal of Consciousness Studies 2002;9(11):3–29.
35. Kim J. Mind in a Physical World. First ed. Cambridge, MA: The MIT Press; 1998.
36. Block N. The harder Problem of Consciousness. The Journal of Philosophy 2002;XCIX (8):1–35.
37. Hardcastle VG. On the Matter of Minds and Mental Causation. Philosophy and Phenomenological Research 1998;58(1):1–25.
38. Churchland PS, Sejnowski TJ. The computational brain. Cambridge, MA, MIT Press; 1992.
39. Abbott L, Sejnowski TJ. Introduction. In: Abbott L, Sejnowski TJ, editors. Neural Codes and Distributed Representations: Foundations of Neural Computation. 1st. ed. Cambridge, MA: MIT; 1999:vii-xxiii.
40. Koch C, Davis JLE. Large-Scale Neuronal Theories of the Brain. First ed. Cambridge, MA: The MIT Press; 1994.
41. Bickle J, Mandik P. The Philosophy of Neuroscience. The Stanford Encyclopedia of Philosophy, 2002 (http://plato.stanford.edu/archives/win2002/entries/neuroscience/).
42. O'Reilly RC, Munakata Y. Computational Explanations in Cognitive Neuroscience. Understanding the Mind by Simulating the Brain. First ed. Cambridge, MA: The MIT Press; 2000.
43. Coenen AM. Neuronal phenomena associated with vigilance and consciousness: from cellular mechanisms to electroencephalographic patterns. Consciousness & Cognition 1998;7(1):42–53.
44. Newman J. Thalamic contributions to attention and consciousness. Consciousness & Cognition 1995;4(2):172–193.
45. Bauer R. In search of a neuronal signature of consciousness – facts, hypotheses and proposals. [Editorial]. Synthese 2004;141(2):233–245.
46. Block N. Two Neural Correlates of Consciousness. Trends in Cognitive Sciences 2005;9(2).
47. Koch C. The Quest for Consciousness. First ed. Englewood, CO: Roberts and Company Publishers; 2004.
48. Steriade M, Contreras D, Amzica F, Timofeev I. Synchronization of fast (30–40 Hz) spontaneous oscillations in intrathalamic and thalamocortical networks. Journal of Neuroscience 1996;16(8):2788–2808.
49. Evans BM. Sleep, consciousness and the spontaneous and evoked electrical activity of the brain. Is there a cortical integrating mechanism?. Neurophysiologie Clinique 2003;33 (1):1–10.
50. John ER. The neurophysics of consciousness. Brain Research – Brain Research Reviews 2002;39(1):1–28.
51. Zeki S. A vision of the brain. 1st. ed. Oxford: Blackwell Scientific Publications; 1993.
52. Humphreys GW. Conscious visual representations built from multiple binding processes: evidence from neuropsychology. Progress in Brain Research 2003;142:243–55, 2003:-55.

53. Viviani P, Aymoz C. Colour, form, and movement are not perceived simultaneously. Vision Research 2001;41(22):2909–2918.

54. Bechara A, Damasio H, Damasio AR. Role of the amygdala in decision-making. Ann N Y Acad Sci 2003;985:356–369.

55. Damasio AR. The Feeling of What Happens. First ed. New York: Harcourt Brace and Co.; 1999.

56. Blanchard RJ, Blanchard DC. Attack and Defense in Rodents As Ethoexperimental Models for the Study of Emotion. Progress in Neuro-Psychopharmacology & Biological Psychiatry 1989;13:S3-S14.

57. Deacon TW. The Symbolic Species: the co-evolution of language and the brain. First ed. New York, NY: W. W. Norton & Company, Inc.; 1997.

58. Critchley M. The parietal lobes. London: Edward Arnold & Co.; 1953.

59. Sperry RW. Consciousness, personal identity, and the divided brain. In: Benson DF, Zaidel E, editors. The dual brain. New York: The Guilford Press; 1985:11–26.

60. Gazzaniga MS. Brain and conscious experience. Advances in Neurology 1998;77:181–192.

6 The word-grounding problem and the incompleteness of language

Summary Our common phylogenetic endowment enables us to communicate with other humans and some higher animals through sounds or gestures. We also have the abstract capacity to develop a complex symbolic language, but language acquires meaning only when grounded in qualitative experiences. Language can name ineffable experiences, but we cannot transfer their meaning through explanations. Thus, language is incomplete because it is not a closed circular system, and needs to be grounded in independent intuitions, as mathematics does. Propositional knowledge is generally considered the "highest" form of knowledge, which allows us to construct our cultural universe and the world of values in which we live. The consistency of science and our capacity to interact successfully with the world illustrate the usefulness and reliability of our intuitions, language, and knowledge.

6.1 We live in a symbolic universe

Humans share with higher animals the capacity to interpret environmental signals and to communicate with each other, but the capacity to invent and use a variety of abstract symbols to communicate complex ideas is the most exclusive of the human abilities. Philosophers have long recognized that language is a symbolic form of communication that is an essential part of our cultural universe. The physical reality seems to be getting smaller because it is more controllable, whereas our symbolic and cultural environment keeps growing, as indicated by the infinite connectivity through electronic media. Ernest Cassirer (1874–1945) believed that we live in a universe of symbols and that we are *symbolic animals*; he distinguishes between *signals* and *symbols*, which he described as belonging to "two different universes of discourse". For Cassirer, signals are parts of the world that some animals and humans can interpret, whereas symbols are exclusive human creations and part of our world of meaning [1]. Biologists have also recognized that our symbolic capacity sets us apart from the rest of the animal kingdom [2–4]. Animal communication is mostly limited to signals—innate and learned gestures and sounds—that carry information, but it is not rich enough to encode the complex symbolism of human language and culture.[1] Today, we are immersed in a universe of words and symbols, and we cannot conceive human life without language, which has become our most distinctive attribute [1].

Language is an enigma that has fascinated ordinary individuals as much as philosophers, scholars, and scientists since biblical times. It is as complex as it is interesting, and its multiplicity of functions becomes evident when we try to define language in a few words. Communication between individuals is one of the most obvious, but certainly not the only function of language. Our whole civilization and culture are based on the symbolic power of words, ranging from the most trivial activities to the transcendental task of serving as the repository of our civilization, and the vehicle for transferring our cultural advances to future generations. Language provides a highly symbolic shorthand that is critical for manipulating thoughts and ideas that cannot be achieved without words and complex sentences. Although we can think without words and with different kinds of images and memories of experiences, our capacity for highly abstract and symbolic thinking is enormously potentiated by the use of the compact information encoded in words. Thus, the multiple problems encountered in defining language reflect both the complexity of language and the layers of symbolism that are essential to describe the intricacies of our mind and culture.

Since primitive cultures left no written records of their languages, linguists have relied on indirect evidence to reconstruct the past. This evidence comes from a variety of sources, such as cross-cultural studies, the development of Creole and sign languages, the language of infants, and the effects produced by brain diseases. Anthropologists and philosophers during the early twentieth century thought that there was some kind of intrinsic association between language and myth. This concept arose mainly because early studies focused on underdeveloped cultures that explained natural causes and human actions as affected by spirits and supernatural forces. Thus, myths and language were thought to be intrinsically associated [1]. Primitive societies believed that some words had supernatural powers, so they continue to use them, even today, as if these words were magic tools to address the gods or to cast evil spells on those they dislike. The idea of language as a special gift from the gods is also encountered in many unrelated cultures [5], from the most isolated and primitive tribes, to the Bible. Early studies of specific languages were carried out by Mesopotamian, Chinese, and Arabic scholars, who were naturally concerned with the proper use of their own languages. Studies of Sanskrit grammar in

---

[1] Campbell's monkeys of the Tai National Park have six types of calls with specific meaning, and vervet monkeys have different alarm calls for three different types of predators: the martial eagle, leopards, and snakes (Nicholas Wade, New York Times 12/08/09).

India culminated five centuries B.C., but this knowledge did not have any impact in Europe until the end of the eighteenth century [5]. European linguists were first influenced by the Greeks, who had a long cultural tradition, as is evident from the sophistication of their philosophy and drama.

6.1.1 The evolution of language and symbolism

By the end of the Renaissance, the stress changed from theoretical studies to *prescriptive* grammars that were used for teaching purposes; this separated professional linguistics into a formal discipline that evolved independently. From the sixteenth through the eighteenth centuries, scholars were involved in the debate between empiricists and rationalists that permeated the approach to linguistics [5]. The empiricists seemed more attracted toward phonetic studies, universal alphabets, and the formalization of the grammars of the different languages, whereas the rationalists were more inclined to develop philosophical grammars.

Leibniz (1646–1716), a rationalist, was interested in the development of a Universal Language (*lingua characteristica universalis*), which would represent concepts in a foolproof manner, using some kind of ideographic or iconic symbolism that could be understood by everybody. Descartes (1596–1650), perhaps the archetype of rationalism, had previously proposed the creation of a universal mathematics, an incentive for the development of the rationalists' grammars that followed the scholastic approaches. The most outstanding proponents were from the abbey of Port-Royal (France) who proposed an all-encompassing theory of grammar to abstract the general grammar underlying all languages. The grammatical principles of universal validity were thought to originate from the nature of human thought, whereas particular differences were arbitrary and mutable conventions as described by Robins [5]. This is a rationalistic principle implying that *the rules of grammar are intrinsic to our minds*. Grammar was conceived as a deductive science. The common mental structure underlying all languages is an idea that later interested Chomsky [6], but as discussed later, this doctrine did not explain how these rules originated or how they found their way into our minds. Grammars cannot develop independently of the need to carry information about the structures and processes that take place in the world.

6.1.2 Language is a highly sophisticated social instrument

Some linguists believe that language is about *combining words to make sentences;* they marvel at the millions of different sentences that an average person can potentially generate, and by the number of different sentences already stored in the Library of Congress [7]. The wonderful potential that we all have is attributed to a mental grammar, some kind of hypothetical program or generative grammar that allows us to produce a practically infinite number of sentences with a finite number of words. This interpretation of language as a "device to make sentences" cannot be taken seriously, because it misses the essence of language, which is to symbolize, to encode information and to communicate rapidly and effectively in the real world.

There are many approaches to language, and there is nothing objectionable in collecting all the rules that apply to different languages or to search for a Universal Grammar. The problem, however, would be to lose the biological perspective and to believe that language is just a system of rules. The common sense idea that language is a way to communicate and to increase the survival of the social group cannot be buried under the complexity of grammatical rules.

Subsequent portions of this chapter elaborate on the idea that *language is a biological function* which increased the probability of survival of our ancestors and became essential for living in a society that needs to communicate effectively. The infinite number of word combinations and the variety of sentences that we can generate is primarily inherent in the complexity of the physical and cultural world in which we live, not in language itself. Paraphrasing the speculative grammarians, we could say that language is only a *reflecting* mirror, a carrier of information. More precisely however, language is a function of the plastic and evolving brain that achieves its highest expression in humans. The essence of language is its symbolic capacity and its ability to encode information. This implies that the information content or semantics is the most important aspect of any language.

6.2 The word-grounding problem

Most of us have believed at some point that the meaning of all words is explained in the dictionary. Paradoxically, this is true for many of the most difficult words, but the words that refer to qualitative experiences are ineffable and a dictionary cannot explain them, as was discussed in Chap. 5. For example, the dictionary defines "red" as the particular hue produced in human observers by the long wavelengths of the visible electromagnetic radiation, etc. This is a *pseudo* definition because neither a color-blind nor a blind born person could understand the definition nor visualize the color. The word "red" and the corresponding experience (*phenomenal*-red) must be present in our minds beforehand to establish the relationship (see Chap. 5 and Table 5.1). The key issue is that *language alone does not give meaning to words*. The fundamental words are *grounded* in qualitative experiences, which are the *common phylogenetic endowment* that allow us to relate to other humans and to domestic animals. The meaning of language, gestures, and attitudes is grounded in our shared emotions and experiences.

The problem of how words refer is a central issue in the philosophy of language [8] and in several related disciplines. To limit the scope of this discussion, I will focus on how words are grounded in experiences and why they are as ineffable as the experiences that they name. The need to design computers and robots that can communicate with each other and with humans has revived the interest in how words refer, which is related to the problem of *symbol grounding* [9]. Harnad asks, "How can the semantic interpretation of a formal symbol system be made intrinsic to the system, rather than just parasitic on the meanings in our heads?" As he indicates, this problem is related to John Searle's Chinese room argument and to the impossibility of learning Chinese using only a Chinese-Chinese dictionary [10]. The grounding^2 of symbols and words consists in connecting them to human experiences or programming them to the data provided by the sensors in robots. *Understanding* the meaning of words or the intentionality—the aboutness—of speech requires words grounded in experiences.

6.2.1 John Searle's Chinese room argument

Searle's Chinese room argument specifically refers to the incapacity of purely symbolic systems to have intentionality, that is, to bring up the meaning or aboutness of words and complex sentences. In an imaginary experiment, John Searle is given several batches of papers with Chinese writing, which he does not understand [10]. He is also given separate instructions in English about how to correlate the different sets of formal symbols. After some time and assuming that the feat is possible, Searle learns to shuffle the papers to produce correct answers that are indistinguishable from those of native Chinese speakers. Thus, no one can tell by reading the results that Searle really does not speak or write a word of Chinese. Similar manipulations could also be done by a computer. These examples show that manipulating symbols according to the appropriate syntax does not produce understanding. Neither Searle nor the computer understands Chinese. The responses of Searle-in-the-Chinese-Room are like the responses of a perfect Turing machine that can answer questions in a manner that cannot be distinguished from that of a human. The answers are syntactically correct and they make sense to a human observer that understands the language, but Searle-in-the-Chinese-Room and the Turing machine miss the *phenomenal meaning or intentionality* of the symbols, because these symbols are not grounded on their experiences. Thus, both are *parasitic systems*, even if they are computationally correct. Searle-in-the-Chinese-Room and the Turing machine are *closed*

---

^{2} In robotics, grounding has been defined as "the process of creating and maintaining the correspondence between symbols and sensor data that refer to the same physical objects" [11]. See also Vogt [12]. In reference to humans, grounding means the association between words and specific experiences.

undecipherable systems, in which every word is defined by a string of words or symbols, which do not refer outside the system.

Closed (non-grounded) systems—such as computers—cannot understand the meaning implicit in the syntactically correct language that they use, because they cannot refer to experiences, which reside *outside* a (closed) circular lexicon. Harnad equates this to learning Chinese using only a Chinese-Chinese dictionary. Thus, the dictionary should not be illustrated like "My First dictionary" with figures and colors that refer to experiences. Part of the problem, as Harnad [13] also indicates, is to find the "right connection" between the symbols and the real world. He adds that a hybrid approach to grounding might work something like a connection between "symbols" and the "sensori-motor projections of the *kinds* of things that the symbols designate" [13].

The problem of intentionality, however, is that in addition the system must not only be aware of the what-it-is-like, but also able to discover the what-it-is-about of experiences. To provide understanding, the system will require structures that realize all the steps outlined in Table 6.1, including their

Table 6.1 The word-grounding problem and the three virtual lexicons

The following are some of the most important elements of word grounding and cognition.

The *phenomenal-intentional lexicon* is conceived as a virtual language-independent lexicon. The "entries" are the what-it-is-like of experiences, and their "definitions" consist in their aboutness and in their indexical reference in a natural lexicon. This lexicon constitutes the phylogenetic endowment of the species, a Rosetta stone for all cultures. The existence of this lexicon in all the members of the species makes language and language translation possible. Our common phylogenetic endowment also provides the bases for understanding anthropoids and domestic animals and for empathizing with them.

The *verbal-phenomenal lexicon* is the virtual nexus between the phenomenal-intentional lexicon and the regular verbal dictionary. I conceive it as a "bilingual" dictionary that associates our phylogenetic endowment with the signals, utterances and words developed by specific cultures. This lexicon grounds words in the phenomenal, opening the doors between phenomenal and propositional meaning. Word grounding is implemented by giving a name to experiences, phenomenal concepts, or their aboutness. In contrast, robots ground symbols/words in sensor data, which so far remain unconscious and without intentionality.

The third lexicon—the *personal version of the regular dictionary*—explains the meaning of words with definitions. However, some definitions must eventually resort to examples to refer to the meaning of ineffable words, which are anchored in the verbal-phenomenal lexicon. Thus, when I say "red" it seems that I am referring directly to the objective red color, but I am actually referring to the red quale and to the underlying neural processes that ground the word "red" in my brain. Thus, the reference of "red" to the objective red color is indirect through phenomenal red (see Tables 5.1, 5.2).

incorporation into conscious processes. The sensorimotor robotic capacity described by Harnad is not enough to produce experiences (phenomenal understanding). The evidence from neuroscience indicates that, in animals and humans, neural structures in the sensory areas of the brain provide a model or surrogate such as neural-red, which must be incorporated into conscious processes to become phenomenal, i.e., phenomenal-red, and generate experiences (see Chap. 5, Table 5.1).

To understand the aboutness of experiences and to survive in a primitive environment, *we do not need a language*, as shown by higher animals and small children. Dogs and monkeys understand many experiences and situations, which they can handle successfully. However, the understanding of organisms without language does not transcend their limited and concrete world, and they cannot efficiently communicate complex experiences. Language is essential not only to communicate more effectively with similar systems, but also to further enhance the cognitive possibilities of phenomenal knowledge with the combinatorial capacity of language [4]. All these considerations indicate that a closed symbolic system is not sufficient to produce understanding, even if the system could be automated to pass the Turing test. As John Searle indicated, intentionality (i.e., knowing-what-it-is-like) is not produced by symbolic shuffling.

6.2.2 Robots ground words or symbols in sensor data

The symbol-grounding problem was brought about by the need to communicate with robots and by the practical and theoretical interest in how robots might communicate with each other. In trying to solve these problems, we are reversing the natural course of biological evolution in which cognitive systems became conscious and acquired intentionality *before* they evolved a language. Cognitive awareness is independent of language and phylogenetically much older [4]. Thus, developing a way to communicate with unconscious machines creates some special problems. In an oversimplified view, robotic systems include two main types of components: one that reasons about the abstract representation of objects (symbolic reasoning system) and another (sensorimotor) that has access to perceptual data [11]. Grounding is the process of creating and maintaining the connections by which the symbols (such as the name of an object) and sensor data refer to the same physical object. This means that robots should ground words in processes that are homologous to our perceptual analogues or neural surrogates (such as *neural-red* or *neural*-hunger, Table 5.1). The fundamental difference is that, in robots, the perceptual analogues are not incorporated into conscious processes to become experiences.

6.2.3 *Humans ground words in the verbal-phenomenal lexicon*

Humans and other animals learn not only the meaning of their experiences, but also how to interpret attitudes and to guess through behavioral clues the intentions of others; animals can infer phenomenal states (emotions) in other individuals and communicate meaning through signals and attitudes. This has high survival value, but animals without language cannot formulate their thoughts in sentences. In contrast, humans can refer to experiences and their aboutness through signals and words.

Signals, language, and propositional knowledge are possible because we can refer to qualitative experiences, phenomenal concepts, and their aboutness by constructing a virtual *verbal-phenomenal* lexicon (Table 6.1). For example, when I say: "These peaches are sweet", you understand what I mean because we both speak English and "sweet" refers to the taste experience (*phenomenal*-sweet) produced by sugar, honey, molasses, ripe fruit, etc., which we have both previously experienced. You learn something new about *these* peaches, but you must know in advance the what-it-is-like of tasting something sweet (the *phenomenal concept sweet*) to understand the proposition. We can name and then we can discuss what we both already know through ineffable experiences, providing that we have enough circumstantial evidence to believe that we are talking about the same "you-know-what-I-mean" (i.e., the what-it-is-like of *phenomenal*-sweet). However, there is no way to prove that homologous experiences or phenomenal concepts are identical in different individuals. Thus, it is more correct to say that words only *connote* the phenomenal experiences in which they are grounded. We can refer to experiencing red color, sweet taste, or anxiety, even if we cannot actually communicate the corresponding phenomenal concepts through propositional explanations.

The neural structures that realize experiences are genetically determined. Moreover, the anatomical and physiological uniformity of the brain structures that realize homologous experiences suggests that their basic phenomenal qualities are similar in most normal humans. This implies that *the species has a common endowment* of potentially similar qualitative experiences that could serve as the bases for understanding between the members of the same species and for grounding the referring signals or words, which then become meaningful to the group. On the other hand, we might never be able to understand extraterrestrials, because the qualitative character of their experiences and phenomenal concepts would be contingent upon the senses and neural structures developed through their extraterrestrial evolution. That is to say that the relation between an objective stimulus and the corresponding qualitative experience is contingent upon the structure of their brains. These brain structures could be differently implemented in different animal species. However, the relationship between the neural surrogates elicited by a perception and the qualitative experience is that of an identity, which requires no further explanation [14].[3]

In contrast to *phenomenal* reference (intentionality), language or *propositional reference* uses signals or words to refer to qualitative experiences and to their aboutness. Our ability for naming or referring to qualitative experiences and phenomenal concepts provides the bases for grounding words. Giving names to ineffable experiences provides the bridge to propositional knowledge and solves the problem of *language grounding*,[4] which is analogous to, but much more complex than the problem of symbol grounding [9].

In humans, word or symbol grounding takes place by association and memory of experiences, so propositions can refer to experiences simply by *naming* them. Thus, words and propositions refer to the objects and properties of the external world indirectly. This means that phenomenal concepts (e.g., knowing the what-it-is-like of being hungry) are an *absolute* requirement for grounding not only the aboutness of a specific experience (phenomenal reference or biological meaning, e.g., hunger is about eating), but also words—the symbolic referent of language and propositional knowledge. The neural surrogates (Table 6.1) of phenomenal concepts (knowing the what-it-is-like) and their aboutness are essential to grounding the meaning of words that refer to them. Thus, for a language to have intentionality some of its words must be grounded on the what-it-is-like and on the what-it-is-about of qualitative experiences.

In summary, we ground words by constructing the verbal-phenomenal lexicon that serves as a "bilingual" dictionary necessary for establishing the correspondence between the neural processes that realize experiences (phenomenal hunger) and their aboutness (eating) and words that refer to them. The phenomenal reference provides *grounded* words with meaning and avoids the circularity and the lack of intentionality of a closed language (which has no ineffable words because every word is defined by a set of words or symbols). The meaning of ineffable words consists in phenomenal experiences or *conscious neural processes*, which cannot be duplicated in other brains through the symbolism of language. Without phenomenal grounding, a language is closed, impenetrable, and parasitic. As illustrated by Searle's Chinese room argument [10], a closed system of symbols cannot refer to experiences, so it does not provide intentionality. Moreover, a closed symbolic system cannot be decoded. Grounding words in experiences avoids the meaningless circularity of a language without ineffability. All these considerations imply that *the intentionality*

---

[3] The contingency of the relation between qualia in humans (i.e., experiencing something hot or *phenomenal* heat) and their objective reference (mean kinetic energy or *objective* heat) has been discussed by Kripke [15, 16] and by Papineau [14, 17].

[4] Gibbs has expressed a similar view [18] in which linguistic meaning is embodied in experience [19], giving empirical evidence for a relationship between meaning of sentences and human action.

of language is borrowed from phenomenal experiences by grounding words through a second dictionary, the *verbal-phenomenal lexicon*, which is analogous to a bilingual dictionary (Table 6.1). The association between words and experiences consists in a physical association between their neural surrogates.

6.3 The incompleteness of language

The "incompleteness" qualifier suggests that language is somehow deficient, but that is not the case. Language is incomplete in the same sense that mathematics is incomplete. Incompleteness refers to our incapacity to define every word or symbol without eventually referring to qualitative experiences. Actually, the need to refer to experiences and intuitions—the core of our subjectivity—is the most important property of language, without which we would not be able to communicate. In this respect, language is similar to mathematics, because both are grounded on intuitions.

A highly informative coverage of the incompleteness of mathematics can be found in Rebecca Goldstein's book *Incompleteness: The Proof and Paradox of Kurt Gödel* [20]. This fascinating book captured my imagination and was hard to put down, even though following the mathematics behind Gödel's proof was not easy reading. Ms. Goldstein has a Ph.D. in philosophy and is a gifted novelist; she describes Gödel's discovery of the incompleteness of arithmetic, his ensuing fame, and his personal incapacity to relate to people and to conduct his life. However, when the importance of his discovery was recognized, he was invited to move to Princeton, where he became a good friend of Albert Einstein. He and Einstein used to take long walks through the campus and they seemed to have a deep admiration for each other's theories.

The incompleteness of arithmetic was shown by its inability to justify itself.[5] This means that any self-contained closed system of symbols is not demonstrable from within the system. Actually, closed symbolic systems, which are those that are not grounded in experiences, are incomprehensible to outsiders because their symbols can neither be decoded nor translated. As previously discussed, it would be like learning Chinese from a Chinese-Chinese dictionary. Gödel's proof had wide implications and invalidated the idea of the self-sufficiency of mathematics, from the axiomatic geometry of Euclid to the formal system that Hilbert tried to develop early in the 1900s.

Gödel's results also cast doubts on the completeness of the abstract world of Platonic forms, since Plato considered that all *forms* were part of an *ideal realm*

---

[5] Gödel's results have been corroborated by Nagel and Newman [21] and by generations of students and teachers; they all agree that his results show the incompleteness of the formalism in mathematics and rule-based computational systems.

that must be discovered through reason. Gödel actually contradicted himself, because he never renounced his Platonism, even though his demonstration implied that the world of abstract forms was based on human intuitions, and not on formal expressions. This is not surprising given that Gödel was a dualist who believed humans have souls and denied the theory of evolution [20]. Paradoxically, the most outstanding logician of the twentieth century lived in an imaginary world, full of contradictions. Unfortunately, he was paranoid throughout most of his life. At one point, he was terrorized by his refrigerator, which he thought was emitting poisonous gases. Gödel died in 1978 of self-starvation and malnutrition caused by his fear of being poisoned. However, at the time of his death, Gödel was about to publish an article in which he showed that, by *appealing to intuition*—by accepting certain assumptions about mathematical reality—he could then demonstrate the consistency of an arithmetic system [20]. Thus, to *ground* mathematics on solid bases, it is necessary to appeal to intuitions, which grasp the existence of a reality that lies *outside* the symbolic system. In other words, we need to find an intuitively understandable link, such as a relation of a number or a word, to decode any language or symbolic system.

On reading *"Incompleteness"*, the striking similarities between the incompleteness of mathematics and language become evident. Language is incomplete in the sense that we cannot explain the words that *name* qualitative experiences. The ineffability of qualia and of the words that name them are vexing problems that have preoccupied some philosophers for years. However, my fascination with the wonders of language emerged not from philosophy, but from observing how rapidly children learn a language and how easily it can be destroyed in an adult after a small cerebral vascular accident. We cannot avoid recognizing our fragility when we see how quickly we can lose our most distinctive attribute. When dealing with patients with aphasias and other disorders of language, one comes to realize the overwhelming complexities of language that are evident behind the individual symptoms.

As a student, my interest in language was rekindled when I became aware of the impossibility of explaining qualitative experiences through language. Even though some philosophers of mind believe that ineffability lies in the *mysteriousness of qualia,* I was not convinced. In fact, *qualitative experiences preceded the development of language* by millions of years. Since Darwin, we have known that animals can express their emotions and understand each other, at least at a basic biological level, without using a verbal language. Moreover, observing the essential role that experiences and emotions have in animal survival and communication, there can be no doubt that higher animals have developed some form of "folk" psychology that allows them to understand each other, even across species and in the absence of a verbal language. The mutual understanding and emotional attachment that people develop with dogs and other domestic animals clearly indicate that some primitive form of basic understanding and a common body language must have preceded the development of verbal languages.

The ineffability of experiences does not have to be explained once we recognize that experiences consist in neurophysiological models or neural processes that acquire their own meaning in a *language-independent* fashion by association with other experiences. Despite the ineffability and the language-independence of experiences, experiences are the "doors" to perceptions and understanding. The aboutness of experiences is naturally established by association with other experiences. Thus, the what-it-is-like and the what-it-is-about of experiences are empirically established, even in the absence of a verbal language. This implies that the most basic *preverbal* dictionary that we have is a virtual language-independent *phenomenal-intentional lexicon,* in which the "entries" are the what-it-is-like of experiences and their "definitions" consist in their aboutness. Accordingly, *hunger* is about the need to eat, *thirst* is about the need to drink water, and *pain* is something unpleasant to be avoided.

One of the most curious and somehow misleading characteristics of experiences is that they are *transparent*, which means that their brain neurophysiological realizers are imperceptible as such; we only perceive the result, the experiences. In consequence, experiences *seem non-physical*, so they are considered by some to be of "mental" or "psychological" nature, which in most minds implies something non-physical, and this fuels people's belief in a supernatural soul. However, the empirical evidence gathered by neuroscience shows that phenomenal experiences are physical processes that are realized by the simultaneous participation of several areas of the brain. Thus, being human ultimately resides in our common experiences, which serve to ground words and make language possible. We absolutely depend on the language-independent *phenomenal-intentional lexicon* (Table 6.1), in which the "entries" are the what-it-is-like of experiences (hunger) and their "definitions" consist in their aboutness (eating).

6.4 Propositional knowledge and language

Information on facts, data, or experiences is expressed as verbal propositions, which are the forms in which knowledge can be stored and exchanged. Thus, propositional or declarative knowledge is also known as explicit or factual knowledge. Propositional knowledge is a distributed function—a collection of neural processes—implemented mainly through *phylogenetically new* areas of the brain, which are highly developed in humans, as discussed in Chap. 3. Propositional knowledge is also generally considered the "highest" form of knowledge because it requires language, the most characteristic human attribute. Propositional knowledge is the type of knowledge with which philosophers are mostly concerned; it is the core of epistemology, and it has to satisfy certain conditions to be considered knowledge. For example, knowledge should be true, and not a simple belief. In contrast to phenomenal knowledge, which consists in ineffable experiences realized by neural processes, *propositional knowledge* is highly symbolic and expressible in different natural and artificial languages. Another essential characteristic of propositional knowledge is the capacity to *name* phenomenal experiences, objects, processes, properties, etc. This is true even though the name of an experience may not always mean the same to everyone. Scientific knowledge can be expressed in propositions, but propositions cannot duplicate either the neural processes that constitute qualitative experiences or the phenomenal concepts that experiences provide.

In contrast to the *hard-wired* ineffable experiences that serve as the bases for phenomenal knowledge, propositional knowledge is *highly symbolic* (Table 5.2), allowing for the storage and transmission of information in several layers of symbolism, from different languages to myth and science [1]. Propositional knowledge is somehow equivalent to "knowing that".[6] This knowledge can be formalized into a coherent body of statements which allow for communication and for development of concepts and theories. We communicate knowledge through language, but language has a much broader role. Language is a way to generate an infinite variety of representations, and provides an inferential engine for predicting, organizing, and planning [4]. In addition, the symbolism of natural and artificial languages allows us to use instruments (from the abacus to electronic computers) that accelerate and potentiate our thought processes.

Propositional knowledge is rooted in basic cognitive elements that can only be provided by phenomenal experiences and concepts. Thus, propositional knowledge, no matter how symbolic, must eventually refer to ineffable experiences and phenomenal knowledge. Giving names to phenomenal concepts provides the bridge between phenomenal and propositional knowledge. There are several but finite degrees of separation between the words and symbols used in propositions and the neural equivalences provided by phenomenal experiences and by phenomenal concepts. The layers of symbolism in propositional knowledge seem like Russian dolls, because they give us the initial impression that the series continues forever. The same can be said for the chain of symbolic references implicit in words. However, after we find the last "doll", we have to ask, "Do you know what I mean?" which is a reference to an implicit, ineffable phenomenal concept that can be acquired only through qualitative experiences. For example, when I say, "the ripe tomato is red", I am

---

[6] For reasons of space, I will not be able to discuss the traditional problems of either epistemology or linguistic semantics. I will refer to a proposition in a non-technical fashion, as "A statement or assertion, consisting of a predicate and a set of arguments" [7]. Thus, in the most general sense, propositions consist of information that describes or represents the world. Propositional knowledge is the knowledge expressed and acquired through a natural language.

not referring directly to the color of the tomato (which does not have any) but to the qualitative experience of seeing red, which results from a brain process triggered by the light reflectance properties of the ripe tomato (Table 5.1).

The anatomical and physiological uniformity of the (homologous) brain structures that realize experiences suggests that the basic qualities perceived are similar in most normal humans. The contingency of the relation between qualitative experiences (i.e., the *phenomenal* heat sensation) and the objective reference (mean kinetic energy or *objective* heat) have been discussed elsewhere [15, 17]. The contingency of how the properties of the external world are experienced, as shown by synesthesias, have been known to neurologists for many years [22, 23]. Thus, from any angle that we look at it, the relationship between experiences and objective reference is contingent. Most likely, the sense organs and the brains developing in different worlds would also be different. This is in fact true even in our existing world, where some animals having different sensors and brain structures, such as bats, fishes, and ground moles, must have some qualitative experiences that are different from our experiences.

Words can evoke the phenomenal concept of what-it-is-like to eat chocolate, but words cannot rekindle past experiences. The memory of an experience must be distinguished from the experience itself. The ability to *reenact* experiences from memory would imply *being able to reverse the normal flow of information* to activate the brain areas that realize the neural equivalents of specific qualitative experiences. Aside from its physical impossibility, given the structure of the brain, such *reverse cognitive ability* would generate a serious biological disadvantage, in allowing spurious memory-evoked satisfaction of functions essential for the survival of the species, such as eating and mating. Animals that could satisfy their hunger by evoking the memories and satiation from their previous meals would not survive.

Propositional knowledge, despite its higher order and complexity, can neither mimic qualitative experiences nor activate the primary sensory areas of the brain to generate them. The brain structures that implement propositional knowledge cannot model the activity of the phylogenetically older areas, which have a highly specialized structure that evolved to generate specific qualitative experiences. The fundamental differences between phenomenal and propositional knowledge[7] are summarized in Table 5.2. The machinery of

---

[7] Propositional knowledge is encoded in language, of which words are its basic element and first organizing principle. Words (graphemes) are assigned a conventional meaning that constitutes the mental dictionary or lexicon, which may be organized in the brain according to categories. Evolutionary considerations suggest that primeval words, like the sounds emitted by animals, effectively communicate emotional states or phenomenal concepts. The second organizing principle of language is grammar, which has three components: morphology, syntax, and phonology [7].

language appears to have been designed to encode and decode information for sharing it with others [24].

6.5 Summary and conclusions

Both arithmetic and language are incomplete open symbolic systems containing processes that cannot be defined but are intuitively understandable and essential for communication. Thus, all words must be directly or indirectly grounded in our experiences (the what-it-is-like) which are associated to other experiences that provide their meaning (aboutness). The impossibility of justifying ethics and esthetics by purely abstract principles (or axioms) is also similar to the impossibility of justifying a closed mathematical system. This is because our *intuitions* and *experiences* necessarily lie at the core of everything that we can understand. Experiences are the link that gives unity to our symbolic universe. However, this also means that we live in some kind of Platonic cave which is difficult or impossible to transcend. Our genetic endowment and experiences provide all possible meaning to what there is, or to what we can understand. We are complex systems with experiences and symbolic languages which together form a closed system that presumably precludes any emotional interaction with extraterrestrials. The symbolic forms that characterize our culture—such as mathematics, logic, language, ethics, and esthetics—are grounded in our common experiences. This implies, as Rebecca Goldstein has said, that we might be "the measure of all things", as originally proposed by the Sophists in the fifth century B.C.

References

1. Cassirer E. An Essay on Man. An introduction to a philosophy of human culture. 1st ed. New Haven and London: Yale University Press; 1944.
2. Goldstein K. Human Nature in the Light of Psychopathology. 1st ed. Cambridge, MA: Harvard University Press; 1951.
3. Goldstein K. The Organism. A Holistic Approach to Biology. New York: American Book Company; 1939.
4. Deacon TW. The Symbolic Species: The Co-Evolution of Language and the Brain. First edition ed. New York, NY: W. W. Norton & Company, Inc.; 1997.
5. Robins RH. A Short History of Linguistics. 2nd ed. London and New York: Longman; 1979.
6. Harris RA. The linguistic wars. 1st. ed. New York: Oxford University Press; 1993.
7. Pinker S. The Language Instinct. 1st. ed. New York: William Morrow and Co., Inc.; 1994.
8. Evans G. The Varieties of Reference. Oxford: Clarendon Press; 1996.
9. Harnad S. The symbol grounding problem. Physica D 1990;42:335–346.
10. Searle JR. Minds, brains, and programs. Behavioral & Brain Sciences 1980;3(3):417–457.
11. Coradeschi S, Saffiotti A. An introduction to the anchoring problem. Robotics and Autonomous Systems 2003;43(2–3):85–96.
12. Vogt P. Anchoring of semiotic symbols. Robotics and Autonomous Systems 2003;43 (2–3):109–120.

13. Harnad S. Symbol Grounding and the Origin of Language. Computationalism: New Directions . Cambridge MA: MIT Press; 2002:143–158.
14. Papineau D. Mind the gap. In: Tomberling JE, editor. Philosophical Perspectives, 12, Language, Mind and Ontology. First ed. Malden, MA: Blackwell Publishers, Inc.; 1998:374–388.
15. Kripke SA. Excerpt from "Identity and Necessity". In: Block N, editor. Readings in Philosophy of Psychology. 1st ed. Cambridge, MA: Harvard University Press; 1980:144–147.
16. Kripke SA. The Identity Thesis. In: Block N, Flanagan O, Guzeldere G, editors. The Nature of Consciousness. 1st ed. Cambridge, MA: The MIT Press; 1997:445–450.
17. Papineau D. Thinking About Consciousness. New York: Oxford University Press; 2002.
18. Gibbs RW, Jr. Embodied experience and linguistic meaning. Brain & Language 2003;84 (1):1–15.
19. Glenberg AM, Kaschak MP. Grounding language in action. Psychonomic Bulletin & Review 2002;9(3):558–565.
20. Goldstein R. Incompleteness. The Proof and Paradox of Kurt Gödel. 1st ed. New York— London: W. W. Norton & Company; 2005.
21. Nagel E, Newman JR. Gödel's Proof. First, 14th printing ed. New York and London: New York University Press; 1986.
22. Cytowic RE. The Man Who Tasted Shapes. 1st ed. New York: G. P. Putnam's Sons; 1993.
23. Ramachandran VS, Hubbard EM. Synaesthesia—A window into perception, thought and language. Journal of Consciousness Studies 2001;8(12):3–34.
24. Pinker S. Language as a psychological adaptation. Ciba Found Symp 1997;208:162–172.

7 The roots of knowledge

Philosophy seems at present to be in a stage of transition between the a priorism
of the past and perhaps an experimental philosophy in the future.
—*K. J. W. Craik,* The Nature of Explanation 1942.

Summary Most knowledge requires complex forms of learning and several types of memory that serve different functions, localized in different parts of the brain. What we commonly call memory refers to the declarative or explicit memory for language. This memory is the core of propositional knowledge, mainly localized in the left frontal and temporal lobes of the brain; it is affected early on by Alzheimer's disease, which damages the hippocampus, a structure that plays an essential role in memory consolidation. In contrast, the memory for sensorimotor skills, such as cycling or swimming, is language-independent and involves the extrapyramidal system affected by Parkinson's disease. The physical encoding of different types of memory in distinct brain circuits contradicts the popular belief that knowledge is a non-physical psychological process.

7.1 The neurobiology of knowledge

Sensing prey and predators within the environment is essential for the survival of humans as well as other animals that are dependent on the organic materials which are available only from plants or other animals. Most plants can also sense their environment, as shown by their leaves turning toward light and their roots toward the soil and water, but with the exception of a few insect-eating plants, they cannot actively trap food or avoid predators. Plants, bacteria, and the less intelligent animal species survive mainly by having a large number of descendants. Some bacteria and other unicellular organisms sense

J.M. Musacchio, *Contradictions: Neuroscience and Religion,*
Springer Praxis Books, Popular Science, DOI 10.1007/978-3-642-27198-4_7,
© Springer-Verlag Berlin Heidelberg 2012

their environment, detect food sources, and *automatically* move in their direction to internalize food without awareness. In contrast to bacteria, we do not internalize the objects that we sense in the process of sensing them, so the objects that we perceive never get *directly* in contact with our brains. Aristotle (384–322 B.C.) already knew that "the sense is the recipient of the perceived forms, without their matter" [1]. Today, we recognize that *nerve signals*, which are composed of multiple action potentials (Fig. 5.2), are the only connection between the external world and the brain. Nerve signals are triggered by different forms of energy, such as light, air vibrations, mechanical pressure, or odoriferous molecules, which stimulate specific receptors in sensory organs.

Sensing the environment is not enough to produce what philosophers call *knowledge*, a term that they reserve for the higher forms of awareness and understanding found only in humans. However, the survival of many other species leaves no doubts that they have some knowledge of their environment. Chimpanzees and other anthropoids are known to learn from experiences, to make choices, and to a limited extent to predict the behavior of their kin and enemies. The young learn from their parents and peers how to find food and recognize their territories and mates, as well as the pecking order and the social structure of the group [2]. The capacity to learn about the environment and to act in an intelligent fashion gives humans and some other animals a distinct superiority over plants and over less intelligent animals. Higher intellectual abilities, such as better memory, learning, and reasoning, further increase the capacity of humans to survive and to control other species.

7.2 Memory is an essential property of the nervous system

The memory for events and experiences is the type of memory that is commonly studied by psychologists and discussed by philosophers. However, *memory is a much more general property of biological systems*, which is found at many levels of organization, from the whole animal to the cellular and subcellular structures. If memory is conceived as a trace left by an experience, or as a change produced by a physiological process, then memory is one of the most fundamental properties of life [3]. Indeed, different kinds of memory are now known to be crucial for such seemingly diverse processes as immunity, allergic reactions, and drug addiction.

Memory and learning are related functions that have been studied in different systems, from normal humans to flies and unicellular organisms. The genetic endowment is not directly affected by individual experiences, but memory and learning undoubtedly contribute to individual survival and to the passage of the most advantageous genes to the next generation. Bertrand Russell wrote in 1928, "The most essential characteristic of mind is memory" and added, "memory is clearly connected with a certain kind of brain structure, and since this structure decays at death, there is every reason to suppose that memory must also cease" [4]. Today, there is much more evidence to back up what Russell said. Perhaps because of the increasing familiarity with computers, almost everyone now understands that memory is a physical process. Brain lesions produced by vascular accidents, tumors, or Alzheimer's disease leave no doubt that memory consists in the storage of physically encoded information in the brain.

Even if memory and learning are general properties of biological systems, the central nervous system is specialized in storing and retrieving different types of memory, some in exquisite detail. This is illustrated by the wide variety of methods that have been used in their study, which include psychological, behavioral, electrophysiological, and molecular techniques. Today, it seems trivial to say that information cannot be stored without a modification of a physical medium, but the remark is necessary because memory has been considered classically as a mental or psychological process, which somehow implied that memory should not be considered a purely physical process. Furthermore, *information is a physical kind*, so memory must be *physically* encoded in media that may vary from cuneiform depressions in Babylonian clay tablets to magnetic or optical changes in a memory stick or in a computer hard drive. The brain and other biological systems also store information in several ways, from DNA which encodes genetic information [5, 6] to rapid neurotransmitter-induced changes in the activity of nerve cells [7]. Together, many lines of evidence indicate that information can only be encoded and decoded in physical systems. The simplest forms of learning, habituation, and sensitization have been used as paradigms to study the cellular foundations of learning and memory.[1] In *habituation*, animals and humans become desensitized to neutral, non-noxious stimuli. For example, we stop feeling a wristwatch, spectacles, or shoes after wearing them for a while. Experiments using *Aplysia californica*, a sea snail, indicate that habituation is implemented by a temporary decrease in the chemical signal of the sensory neuron and the excitatory interneuron (Fig. 7.1).

These changes result in a weaker stimulation of the motor neuron and a decreased withdrawal reflex. However, if the training session is repeated a few times within a certain period, the transient changes become enduring, for at least 3 weeks. The enduring memory for desensitization is mediated by a temporary decrease in the *number of contacts of the reflex arch*, which consists in the connections between a sensory and a motor neuron [3]. *Sensitization* is the opposite of habituation, but it is a more complex process in which a potentially harmful stimulus *increases* the magnitude of the next reflex response. Studies in the sea snail have shown that a single aversive stimulation can facilitate the reflex withdrawal to a second stimulation, even if the second stimulation is not

---

[1] See Kandel [8, 46] and "In Search of Memory" [3].

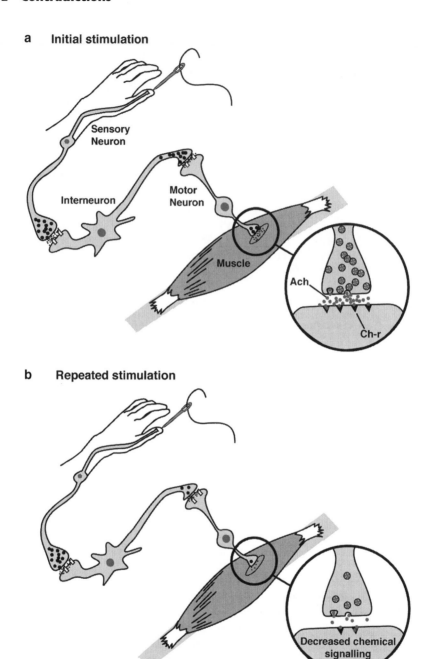

a   Initial stimulation

Sensory
Neuron

Interneuron

Motor
Neuron

Muscle

Ach

Ch-r

b   Repeated stimulation

Decreased chemical
signalling

Habituation

DCR

Fig. 7.1 Sensory habituation. Simplified drawing to show an example of sensory habituation that reflects a decrease in chemical signaling from a motor neuron to a target muscle when the initial sensory stimulus (a needle prick on the thumb) is followed by repeated stimulation (**b**). The small circled regions in (**a**), and (**b**) are enlarged to show

harmful. The biochemical mechanisms that mediate these increases in transmitter release are quite complex. Let us say only that the changes in the short-term sensitization last only minutes, whereas the long-term sensitization that is obtained by repetitive stimulation, requires the synthesis of new proteins and the growth of new synaptic connections as shown in Fig. 7.2 [3, 8, 9].

Habituation and sensitization are elemental mechanisms of rapid adaptation to the immediate physical environment, but they are not foolproof and are prone to errors. These elemental forms of learning and memory have two essential characteristics. One is that the memory is not encoded in separate specialized brain structures, but the learning is *intrinsic to the neural circuits* that receive sensory inputs and generate a motor response, which may include the brain and spinal cord, the two main components of the central nervous system. Thus, there are no specialized memory banks for the simplest forms of learning; the same circuit that carries out the function *encodes* the memory by *facilitating* the next response. The second key characteristic is that these elemental circuits can learn in *two* modes, in a temporary or *transient* fashion that lasts minutes, and in an *enduring* modality that lasts for days or longer (Fig. 7.1). Studies of memory have clearly shown that learning and knowledge are the consequence of physical changes in the function of and in the contacts between neurons (synapses). These findings have far-reaching implications, because they demonstrate the *physical nature* of what was considered "psychological" up to a few decades ago.

7.3 There are different kinds of memory and learning

The most recognizable type of memory is declarative or explicit memory of the kind that fails when we forget names, faces, or phone numbers (see Table 7.1). In contrast, implicit or nondeclarative memory deals with learning skills and habits, such as walking, eating, or playing tennis. Memories can be selectively lost, indicating that different memories are stored in different parts of the brain. Thus, we could distinguish as many kinds of memories as functions the brain performs! In real life, most memories are complex and are implemented through several neuronal networks.

Fig. 7.1 (continued) the junction between the motor neuron and the muscle. In the larger circles, the transmitter (acetylcholine, Ach) is released to activate cholinergic receptors (Ch-r) on the muscle. In the habituation produced after repeated stimulation, less neuro-transmitter is being released. The motor neuron is linked with the sensory neuron by a series of interneurons, of which only one is shown for simplicity

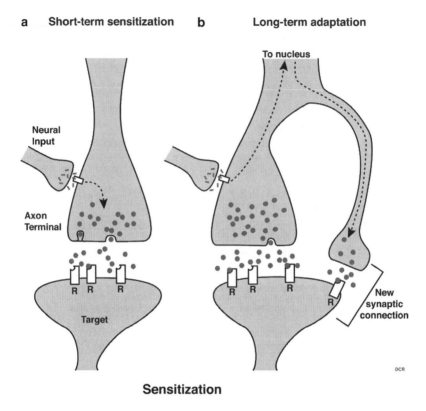

Sensitization

Fig. 7.2 Sensitization and adaptation. Simplified representation of the mechanisms involved in (**a**) short-term sensitization, and (**b**) long-term adaptation. In (**a**), the short-term sensitization is produced by the increase in transmitter release (small filled circles) from the axon terminal onto the receptors (R) on the target cell. In (**b**), long-term adaptation is produced by the growth of an additional synaptic connection. This requires signaling back to the nucleus in the cell body, which in turn results in new protein synthesis and the growth of new synaptic connections

7.3.1 Different types of memory and learning are processed and stored in functionally distinct brain regions

The nervous system has many functions, so it has different mechanisms to store experiences, plans of action, or codes of behavior and these are associated with specific brain circuits. Different types of memory loss can be observed as a result of brain diseases and brain surgery affecting different brain regions. In the late nineteenth century, there were also several descriptions of amnesias; the most striking was Korsakoff's syndrome, a severe form of amnesia with multiple brain lesions that were produced by a combination of alcoholism and malnutrition. More recently, the now classic studies of Penfield and coworkers [10, 11] have provided important information on different kinds of memory and on their specific brain locations (Chap. 4). The recollection of memories induced

Table 7.1   Classification of memory and learning

Memory and learning are generally classified into two main divisions: **(1) declarative or explicit** and **(2) nondeclarative or implicit**.

The **(1) declarative or explicit** memory is damaged in "amnesic" patients with temporal lobe, limbic-diencephalic lesions, including Alzheimer's disease. This memory is encoded knowledge that can be transmitted through declarative statements and recollected at will. This is what is popularly understood as "memory", and is a way in which *propositional knowledge* is encoded. A single experience can be sufficient to leave permanent traces. Declarative memory is associative and "representational" and is used to model the external world. In turn, the declarative or explicit memory can be divided into (1a) *episodic* (personal events) and (1b) *semantic* (new facts).

Declarative memory has also been classified according to its duration. *Short term or working memory* provides temporary storage of conscious information for use in other tasks or for the formation of *long-term memory*. The hippocampal formation in the temporal lobe is essential to consolidate declarative short-term memory into long-term memory, which has several distributed working subsystems: verbal, auditory, visual, spatial, etc.

The **(2) non-declarative or implicit** forms of memory are diverse and include:

(2a) *Procedural memory* for sensorimotor and mental skills and habits. This form is responsible for *know-how knowledge.* Most of what we learn may take place though implicit learning that generates tacit knowledge in an unconscious fashion. Procedural memory includes category learning, which is preserved in amnesia, but is dependent on the integrity of the neostriatum. It also includes abstract knowledge and artificial grammar learning. Implicit learning is considered the *default setting.*

(2b) *Priming and perceptual learning*, which involves interpretation of cues to recall words or objects, is independent of declarative or explicit memory. It is also independent of amnesia and it is related to the neocortex.

(2c) *Associative learning* includes classical and operant or instrumental conditioning, which are functions distributed through several different anatomical systems. The responses consist in emotional, motor and/or autonomic responses, depending on the nature of the conditional stimulus.

(2d) *Non-associative learning* involves habituation and desensitization. The elemental forms of non-associative memory are the most basic, but are not evident to the first person perspective because they are not conscious. These most elemental forms of memory and learning take place at the level of synapses and simple reflex pathways. Different forms of non-associative learning are common to animals and humans. Habituation consists in becoming desensitized to *neutral stimuli*, those without biological value. Sensitization is the opposite phenomenon, in which potentially *harmful stimuli* increase the response to the next stimulus.

---

in some patients by the electrical stimulation of the brain cortex was dramatic, but provided little insight in terms of memory storage sites. The difficulty arose because the stimulation of the same site in the cortex did not always elicit the same response. In addition, the excision of the area stimulated did not necessarily eradicate the memory. These observations gave an early indication that there are *extensive memory networks* that include multiple brain regions. More informative were the studies of the patients in whom the temporal lobes of the brain, including the hippocampus, had been removed because of intractable epilepsy. Brenda Milner, who studied these patients, is credited with the discovery that, to be permanent, human memory requires an intact hippocampus (Fig. 7.3a, b), from where is transferred to several other brain systems [12, 13].

Patients with bilateral removal of the hippocampus, such as Penfield's patient H.M., have intact *long-term memory* for events that occurred *before* the operation, but they *lose the capacity to transfer their short-term (declarative) memory into long-term storage*. These patients can remember names or events for seconds to a few minutes, but once they are distracted, they can never recollect what they apparently knew a few minutes earlier. They present the symptoms of a severe Alzheimer's disease. The hippocampal formation is now well established as being essential for consolidating declarative short-term into long-term memory, even though the precise mechanisms are not fully known. Interestingly, the same hippocampal cells active during the memory-learning period are also active during sleep. This "re-playing" of neuronal activity in hippocampal circuits during sleep is thought to be involved in memory consolidation in rats [14]. The deprivation of REM sleep in humans also has a negative effect on the recall of events from the previous day and on the consolidation of procedural memories [15].

Not all learning is impaired by hippocampal lesions, because patients with such lesions can learn new *motor skills* perfectly well, and repeat their performance when requested to do so. However, they have no awareness or recollection that they have learned a new skill [12, 13]. In contrast, patients with Parkinson's disease affecting the basal ganglia circuits are deficient in learning cognitive skills, but not tasks involving declarative memory [16]. Similarly, learning conditional responses such as the association between the sound of a horn and the approach of a vehicle involves several parts of the brain. The amygdala (Fig. 7.2), which is part of the medial temporal lobe, plays a crucial role in conditional emotional responses [12, 13, 17, 18].

This brief review indicates that the hippocampus and the medial temporal lobe are essential for the formation of *declarative memories*, serving as a temporary storage depot before the short-term or working memory is converted into long-term memory. Working memory has been associated primarily with the bilateral prefrontal and parietal cortical brain regions. In addition, it is known to involve the visual and auditory cortex, as well as other cortical and subcortical brain regions described in Chap. 4 and illustrated in Fig. 4.2. Thus, most of the evidence indicates that working memory, like explicit long-term memory, consists in widely distributed, largely neocortical functions that share the same substrate of broad, partly overlapping neural networks [19, 20]. The medial

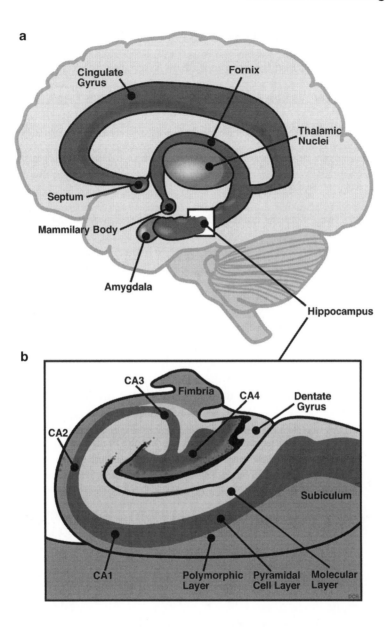

Fig. 7.3   The Limbic circuit. Schematic version of the brain's limbic circuit (dark grey in **a**), which includes the hippocampus, a fundamental structure that is essential to implement long-term memory. The laminar structure and subregions of the hippocampus are shown in the cross-section enlargement in (**b**)

temporal lobes, hippocampal formation, and some adjacent brain structures as well as the medial thalamus are essential for acquiring explicit learning, but not for the recollection of previous explicit knowledge. In contrast to the formation of explicit long-term memory, the capacity to record sensorimotor skills is dependent mainly on the integrity of subcortical brain regions, specifically the neostriatum [21], which is part of the extrapyramidal motor system. Long-term, declarative memory also has several different components, which are mediated by separate brain systems. These forms of memory are intricately linked and can be individually or collectively lost [16, 22, 23].

7.3.2 Declarative memory can be changed into skills and habits

Declarative or explicit memory is readily distinguishable from the implicit memory of sensory-motor skills as revealed by the story of my telegraphy training. When I was a teenager, in the early 1940s, my father decided that, in addition to English, I should learn telegraphy, because he thought that telegraphy would be the way to communicate in the future. He was impressed by the sinking of the Titanic in 1912 and by the communication failure that aggravated the catastrophe, which occurred when he was 11 years old. The chief of the local telegraph office in Lincoln, BA, Argentina, was my father's patient, and a very understanding man, who taught me telegraphy. First, I memorized the Morse code, and then I learned to operate rapidly and with regular movements the electrical switch that sent the signals to the distant receiver. The overall process thus required two steps: first to *learn the equivalence of the two alphabets,* dot and dash = "a", etc., which is an example of *declarative* or *explicit memory*; and second to transmit the Morse signals with rapid movements of the hand and wrist, which requires *learning a sensorimotor skill* and involves procedural or implicit memory.

After a while, I could send messages automatically, without consciously translating the letters into the Morse code. The translation became a fully automatic operation from my eyes to my hand, bypassing the conscious recall of the equivalence between the alphabet and the Morse code. Needless to say, I never used my telegraphy skills. Today, I cannot recall the Morse code, but I can still go through the automatic movements of sending messages. I thus retained the *memory of the sensorimotor skills*, but not the *declarative memory* of the equivalence of the two alphabets. To retrieve the code today, I have to make the hand movements corresponding to a letter and then translate the movements into dots and dashes. This example illustrates that the different kinds of memory are specifically related to the different ways of learning, and to the different types of knowledge. Sensorimotor skills, such as cycling, swimming, and sending telegraphs are not easily forgotten. These differences in memory and learning reflect their physical encoding in different parts of the brain, which were initially recognized from clinical observations showing that the memory for words, faces, music, and eve, for our own body can be selectively lost.

7.3.3 *Memories and images are physically encoded in the brain*

We have the impression that the objects that we see and remember produce internal images in our brain. The vividness of the recollection is variable, but it seems particularly strong for vision during some dreams. In addition, we can visualize faces, hear internal voices, and recollect poems and music. The recollections of past memories can also be produced by electrical stimulation of the brain cortex [11, 24], as previously discussed. However, the perception of mental images is dramatically seen in the hallucinations produced by epilepsy and in schizophrenic patients, who are convinced that they are actually smelling something or hearing real voices.

Neuroscientists and most modern philosophers have objected to the naïve view of looking at "internal images", because this requires the existence of an internal observer, who also requires an additional internal observer, and so on, in a situation that implies infinite regress. Thus, we wonder how is it possible to feel that we can see the internalized information and visualize a face or a building in a way that looks almost "identical" to the corresponding object in the external world. The most reasonable explanation is that both, perceptions and recollections are encoded in a similar fashion. We know—from the scientific perspective—that memories and learning are physically encoded in some of the same neuronal systems involved in their perception.

The encoding of information in gadgets, such as telephones and radios, provides some perspective on the encoding and decoding in the brain. For example, in a telephone conversation, the *air vibrations* produced by spoken words are encoded into electrical impulses by the microphone. Then, the information is transmitted to a central station in the form of electrical impulses or electromagnetic waves that are subsequently sent to the receiver, and this transforms the electrical impulses back into *air vibrations* that mimic the original voice. Thus, the transmission goes through a series of transformations, but ends up *mimicking the original stimuli*, so we do not have any problem in perceiving them. The same is true for the gadgets that store information in CDs and DVDs, which duplicate in a simplified fashion the spatial and temporal cues used by our perceptive machinery.

In contrast to the encoding and decoding in gadgets, the information internalized by the brain *cannot be transformed back into the original input*. This information is encoded as *physical changes in specific neural circuits*, as described previously (Chap. 4). Thus, we never *directly* perceive the objects of the external world as bacteria perceive food. The brain only receives encoded "messages" produced by the senses. Thus, recalling a "mental image" most likely requires some kind of reactivation of the *neuronal changes produced by the original perception*. However, the information that we recall about complex objects is not picture perfect; for example, we cannot count the stripes of the zebra that we saw in the zoo last week. The encoded neural model could be conceived as an analogue spatiotemporal transformation, a gestalt with fewer details than the external object. This means that the only recollections that we can have are *neural surrogates for objects* or encoded signals left by the perception of external objects. We can encode almost anything from an isomorphic analogue (space and shapes) to signals purely made up by the senses and the brain, such as colors, smells, and tastes that are only identified by association with previous experiences (see Chap. 5).

Thus, the "mental" images are not what they seem to us. The memory of a triangle cannot be *literally* triangular. However, we know that *triangularity* must be encoded in the brain, since we can not only imagine, but also draw triangles. The same can be said about straight lines. How can the brain encode a triangle or a straight line if there is nothing triangular or straight inside the brain? While the answer is not known, there are several possibilities. Straight lines and triangles can be described mathematically, so they could be encoded as *analogue processes that mimic the corresponding mathematical equation,* as Kenneth Craik implied [24]. Many biological processes can be described mathematically by engineers; for example, the lines of strength (or force) of certain bones like the human femur were known to follow engineering principles that had been previously described mathematically. This does not mean that mathematics is necessary for the bones to grow, in the same way that mathematics is not necessary to *prescribe* the orbit of the earth around the sun. The earth does not compute its orbit to decide where to move next. Mathematics and the "natural laws" are human descriptions that help us to understand Nature in an analogical fashion.[2]

Some of the spatial and temporal perceptions are isomorphic transformations. An ideal isomorphic transformation would not lose information in the process, whereas qualitative transformations are contingent processes in various approximations, which require additional experiences to acquire their meaning. The relative isomorphism of some perceptions is well illustrated by touch and sight, both of which can sense similar kinds of spatial information. Other forms of perception such as colors are dependent not only on the available light, but also on the light reflectance of the objects and on the anatomical characteristics and physiological condition of the observer (Chap. 5).

---

[2] It is an anthropomorphic interpretation to say that the earth *follows* the natural laws, which consist in approximations made by humans. Natural laws fit reality until they are replaced by better approximations.

Table 7.2 Physical changes associated with memory and learning

A radically reductionist approach is required to find memory traces at the neuronal level. The guiding assumption is that memory is a *physical process.* The approach requires a model system that is simpler and more accessible than the human brain, and the giant marine snail *Aplysia* serves this purpose [46]. The large nerve cells of *Aplysia* are ideal for studying the electrophysiological and biochemical changes induced by learning. This reductionist approach has allowed investigators to characterize the biochemical and structural changes induced in the simple model, and later to look for similar changes in other species and in other, more complex forms of learning.

Neuronal circuits in the hippocampus have the plasticity necessary to sustain explicit memory. This is provided by *long-term potentiation* or LTP, a functional change in neural fibers characterized by a persistent increase in postsynaptic potentials after presynaptic high frequency stimulation. The increase in post-synaptic potentials results in an increase in neurotransmitter release; the potentiation can last for days or weeks, depending on the characteristics of the stimulation. LTP is mediated by activation of transcription factors which results in the synthesis of specific proteins that are essential to implement synaptic potentiation.

The neural code in which explicit memories are written is unfortunately not known. However, the molecular encoding of the simplest forms of learning and memory are now well understood thanks to the ingenious work of several research groups [25–27] and the work of Eric Kandel and coworkers [3, 12, 37, 46] already mentioned. Kandel is a pioneer in using simple systems to provide biochemical and neurophysiological explanations for "psychological" phenomena previously thought to be unapproachable by science (see Table 7.2).

The naïve impression of perceiving things as-what-they-are is an illusion, because sounds, colors, and smells are forms of perception created by our sensory organs and brains. Berkeley [28] was right if we accept the neuro-scientific version of his ideas *as physical processes* that take place in the brain. Most of us cannot easily understand that everything we perceive resides in brain events. The subjective view is that the external world consists in a conglomerate of perceptions with isomorphic (shapes and number of objects) or qualitative encoding (color, pain, heat), which we hope to understand from the scientific perspective. As elaborated in Chap. 5, qualitative experiences are language-independent neuronal processes that enable us to refer to them by their assigned name. However, we cannot communicate the what-it-is-like of experiences through language, as was illustrated by the impossibility of explaining red color to Mary or to a color blind person.

7.4 Implicit learning accounts for the knowledge of skills

The *knowledge of skills or procedures* is a large group of heterogeneous abilities that has two extreme modalities, *sensorimotor know-how* and *mental know-how* [29]. Sensorimotor know-how is essential for human and animal survival and has innate components, such as the suckling reflex that is hard wired in mammals at birth. However, most innate sensorimotor skills require additional learning to become fully functional and efficient, as when fish-catching birds learn to catch fish. They must learn to correct for the distortions produced by light refraction at the air–water interface, the same phenomenon that produces the illusion that a straight stick is bent when it is half submerged in water.

The most fundamental characteristic of the knowledge of skills is that it is a *tacit or language-independent* ability acquired mainly through *implicit, unconscious* learning. Skills improve with instruction and training, but not all the know-how obtained is transmissible through words. Most individuals cannot explain how to maintain balance on a bicycle [30] or how to hit a ball with a bat. Even though all skills are perfected by trial and error correction, tips regarding certain body positions or movements will shorten, but never abolish the learning time. The same can be said for observation of proficient athletes which can make a beginner aware of subtle points that are not evident when concentrating on other details. For example, a violin teacher can explain how to hold and move the bow to reproduce certain notes, and humans and higher animals can improve their skills by watching experts perform. The fundamental point, however, is that the neural circuits involved in the learning processes must be physically modified to develop the coordination between sensory and motor centers. Moreover, certain groups of muscles must also be developed and strengthened by the nerve impulses from the corresponding motor centers.

Mental know-how involves skills that are essential to support the highest forms of learning and knowledge. These forms of learning are the targets of education because they are essential for developing the brain circuits that will be instrumental in acquiring not only language, but also additional mental skills. For example, we learn not only facts, but also mental abilities and logical reasoning.

A considerable part of our learning consists in the *acquisition of information without awareness*, as when we learn grammar [31] and many other "mental" skills (see below). Some philosophers have suggested that know-how knowledge could be reduced to linguistic instructions, but this is not the case. The knowledge of skills must be personally acquired and followed by practice for consolidation. A classic example is the learning by apprentices. "The apprentice unconsciously picks up the rules of the art, including those which are not explicitly known to the master himself" [30]. Arts and crafts provide examples of knowledge that cannot be verbalized, even by their own masters. Thus, know-how knowledge is transmissible mainly by observing the performance of experts, followed by repeated trial and error corrections.

As we know from experience, the acquisition of mental skills, such as learning arithmetic or complex mental operations, also requires years to learn. Knowledge about categories and artificial grammars can be acquired implicitly by accumulating information from multiple examples [32], which include learning causal and logical relationships, grammar, and artificial grammars. Mental skills are fundamental knowledge, because they provide the models and categories for understanding the external world. The capacity to reason, to make moral judgments, and to carry out complex mathematical calculations requires intensive training and time to develop. These abilities also require the development of brain structures able to handle these complex skills. The learning of most complex skills has a *developmental window*, after which, they cannot be fully learned. Certain mental and sensorimotor activities (such as learning calculus or a foreign language without an accent, or playing the violin) are only learned perfectly by young children or adolescents.

Children must also learn how to deal with hunger, thirst, sexual desire, exploratory behavior and other biological needs, which are the *primary engines that generate behavior*. Desires and emotions can be conceptualized as *states of disequilibrium* that must be resolved by an appropriate action, which is naturally the satisfaction of the specific need. Feedback mechanisms check hedonic behavior, suggesting that satiation consists in the restoration of an internal equilibrium. When the behavior results in the satisfaction of the need, it becomes *reinforcing*, which is to say that reinforcement promotes the learning of the satisfactory behavior. Once the behavior is learned, the animal can easily repeat it, whenever the same need arises again. The reinforcement of the learned behavior is an essential characteristic of the process, because the behavior is initially a random or automatic exploratory response. Thus, the positive reinforcement of behavior is one of the key mechanisms for learning, especially in animals and children.

Higher forms of knowledge are a byproduct of not having to fight continuously to obtain food or avoid predators. We become civilized only after our basic animal needs are satisfied. Individuals under life-threatening situations, such as those prevailing during war, famine, or epidemics, tend to revert to and behave as the animals that we are. We can use our brains for purposes that are not essential for survival, such as learning about the classic philosophical questions, only after our biological needs have been satisfied. Bertrand Russell and other philosophers have long realized the cultural value of idleness [33].

In summary, implicit learning and the effects of usage on the development and consolidation of neuronal pathways are essential for learning skills. The implicit and explicit forms of learning are the extremes of a continuum in which we can find complex forms with both components. *Experiences during critical periods are essential to develop the structure and function* of specific brain regions, such as those involved in vision. The utilization of the system develops and programs specific functions and produces a wide range of changes, from the consolidation and development of new synapses to complex structural modifications that take place throughout the nervous system.

7.5 Most knowledge results from complex forms of learning

The forms of learning and knowledge discussed so far are analytical abstractions that are actually components of more complex forms of knowledge (Table 7.1). Most knowledge is the result of a mixture of qualitative experiences and their aboutness, skills, and propositional knowledge, with a component of beliefs and opinions that does not qualify as true knowledge. The knowledge required for the practice of the different professions and crafts is also a mixture of explicit and implicit (know-how) knowledge in variable proportions.

A great deal of the knowledge that we obtain is through observation and imitation. Learning by observation occurs not only in adult humans, but also in children and some other animals. We all tend to imitate practices that look rewarding (positive reinforcement), and avoid those that may result in punishment (negative reinforcement). Modern advertising provides many examples of implicit association between the advertised product and a potential reward, a tempting food or mate, which almost seems to be included in the purchase of the product. Most of the social rules of behavior and fashion are perhaps implicitly learned by observation, and followed by trial and error correction. In contrast to adults, who understand the consequences of the observed acts, children tend to imitate not only other children, but also adults, without understanding the consequences of their actions. It seems as if imitation is part of their exploratory behavior. However, if the results of their imitation are either reinforcing or unpleasant, they will remember them. Older children also may consider aggression acceptable, because it is usually a key to success in television, cartoons, and movies, and unfortunately in some real-life neighborhoods. There are additional forms of knowledge, such as propositional, empirical, a priori, moral, esthetic, etc., which result from multiple abilities, but are too complex to describe in the context of this chapter.

7.6 Neurobiological explanations are reductionist

Biological approaches have been instrumental in explaining our complex body functions, so we also expect science to unveil the mystery of brain functions, including cognition and knowledge. Our capacity to intervene therapeutically in correcting many abnormalities, ranging from major depression to neuropathic pain, certainly indicates that neuroscience yields valuable information about mental health. Neuroscience also offers a reasonable approach for identifying the anatomical and physiological processes involved in the causal chain of events that realize mental functions. Nevertheless, this is easier said than done, and many large gaps in our knowledge remain to be filled. For example, we still do not know exactly all the microphysical neuronal processes required in moving a single finger. We only have a general view of the chain of events required, but the existing evidence has been corroborated by countless studies under normal and pathological conditions and in animal studies.

Scientific explanations are propositional and progressively describe the elements of a function and their causal relations at lower levels of organization and in finer details. This is to say that biological explanations are reductive. In contrast, global explanations are never sufficient, because the constant improvement of our analytical methods allow us to elaborate better explanations by identifying lower and lower levels of microscopic causation. This creates the problem of deciding how reductive an explanation should be to be acceptable.

Explanations in physics and other disciplines are sometimes theory driven, but they must be confirmed by empirical observations. This is true even for the gravitational curvature of light predicted by Einstein. Even so, the success of a single prediction does not prove a far-reaching theory. Moreover, explanations of specific processes must later be incorporated into theories that are more comprehensive. In contrast to physical explanations, most biological explanations are not initially theory driven, because law-like explanations in biology can rarely define complex biological processes.

An observable phenomenon in biology must be empirically explained first by describing its components and then by establishing the chain of causal events that lead to the realization of the function in question. In the past, the most elemental approach was first to disrupt or modify the process of interest using several probes, and then to try to identify their essential components. Hints about their elemental structure could be inferred from the agents that could modify the overall process, like for example the alteration of functions produced by diseases. The preliminary results are then used to generate modest working hypotheses, which are confirmed or rejected. However, we are always surprised, because "Nature" is a tinker that hardly discards anything and often manages to reuse old mechanisms as backup systems or for different purposes [34, 35]. Thus, biological systems are always much more complex and redundant than anticipated, so initial predictions rarely turn out to be complete.

Ideally, scientific explanations are reductive [36] and are initially limited to a single set of problems. As our knowledge progresses, however, we require explanations that go from the apparent, to lower and lower levels of organization that are not perceptible without special tools. In biology, this means adopting a reductive strategy, from whole populations and organisms, to physiological, biochemical, and physicochemical processes. The ultimate reductionist approach may eventually lead to an apparently "different" discipline, in which we have no expertise, such as determining the structure of protein molecules, atoms, and subatomic particles, including their quantum mechanical interactions. The limited expertise of each individual makes collaborations, meetings, and publications essential, not only for confirmation and for general acceptance of the findings, but also for the integration and understanding of complex processes and functions. Theoretical breakthroughs are not everyday occurrences, and most of the time, there is no surprising Eureka moment!

Oversimplified explanations do not work; for example, eliminative materialism has somehow misinterpreted the everyday implications of scientific explanations. Complex qualitative experiences such as love, hopes, intentions, and fears are firmly established by direct incorporation into conscious processes and association to other experiences. Thus, we know that fear or pain cannot be eliminated from our experiences just because we know the brain circuitry and underlying neurophysiological mechanisms. However, knowing their causal mechanistic explanations allows us to develop tools and drugs for minimizing unpleasant experiences and enhancing those that are rewarding; the empirical success of our interventions validates this approach. In many cases, we are able to provide explanations at several levels of organization, but as indicated by Bickle and Mandik [44], we are still missing some of the "glue" that binds these levels together. Thereby, we need to avoid parochial approaches and look outside our "disciplines" for analogies and isomorphic relations if we are to find the bridges between the different levels or generate much more general theories.

Despite the many unresolved questions, science has made enormous theoretical and practical progress during the last century, and the physicality of experiences can be verified by pharmacological, surgical, or electrical manipulations of the brain structures in which these processes are realized. Moreover, the discovery of additional neurochemical processes in these structures allows us to make additional predictions and to intervene therapeutically to modify or suppress unwanted experiences. This strategy has been proven to be effective in modifying pain, and in inhibiting pathological anxiety and panic attacks, as well as partially controlling endogenous depression. Even without complete explanations, experiences can be identified with certain neural processes which can be detected or disrupted by many different physical or pharmacological procedures (fMRI, electroencephalogram, event-related potential (ERP), magneto-encephalography, administration of different drugs, transcranial magnetic stimulation, etc.). Thus, while explaining experiences is in its infancy, everything indicates that we are following the right track, although the road ahead may be long and have many detours.

Complex brain states at a given specific time are as unknowable as the position and velocities of each individual molecule in a gas. Actually, the brain is infinitely more complex and organized than subatomic particles or gases. This should not be a deterrent, because by analogy we should be able to elaborate theories that will help us to understand complex systems. The brain is estimated to have about 10 [11] neurons or nerve cells which are classified in approximately one thousand different types: [37] they form complex networks, with anything from 10,000 to 150,000 contacts per cell. Each neuron also has a complex internal structure, with many organelles and compartments, thousands of different proteins and enzymes that form channels for different ions, transporters and receptors for hundreds of different neurotransmitters, neuroregulators, and hormones. Moreover, the complexity of the electrical properties of neurons and neuronal networks is staggering. Given this complexity, it is easy to understand that complex systems cannot be known in full detail at any given time [38, 39].

Heisenberg's uncertainty principle should be extended to the dynamics of the brain and the multiplicity of its components. Neuroscientists have recognized but rarely stated the impossibility of determining the structure and the state of all the molecules and constituents of a single neuron at any given time. Detecting one component (for example an enzyme or receptor out of two or three thousands) usually alters the overall structure and chemical composition of the neuron, so the analysis of more than one or two additional components is impossible. Thus, our knowledge of the brain is a kind of spatiotemporal collage that we can only understand in principle. The complexity of the brain is such that we cannot obtain an instant picture of all the components of an experience at any specific time.

7.7 Summary and conclusions

Explanations, even those that refer to different levels of organization, should be related to higher and lower levels of organization, but obviously their conclusions cannot be contradictory. Another major, but rarely stressed characteristic is that explanations are propositional, so they are not accessible as experiences in the first person perspective.

There are no full explanations of complex mental phenomena, but many of the processes involved are well known. For example, we know a lot about the neural circuits and neurotransmitters involved in pain perception and about the effectiveness of painkillers, local anesthetics, and opiate analgesics. Similarly, we know a lot about memory and reward circuits, as well as the circuits and neurotransmitters involved in emotional states. Thus, there are no doubts that eventually we will be able to explain qualitative experiences in neurobiological terms.

Propositional explanations are relevant if we can verify some of the steps involved empirically by manipulating the system for therapeutic purposes. However, descriptive empirical explanations can never satisfy our hopes and yearnings. The most that we can do is to establish the identity between an experience and a complex series of neurobiological processes. Once we have checked an identity empirically, we do not need additional explanations, as convincingly argued by Papineau [40]. The problem is that some phenomeno-logically oriented philosophers expect neurobiological explanations to provide the feelings of the experiences explained.

The need for more general theories is obvious, but actually it is hard to formulate them for processes that are not completely known. Thus, modeling the activity of brain functions at many different levels and at different times is one of the most viable options. There have been several successful in silico attempts to devise models of neural networks that could duplicate some functions. These are heuristic approaches that have been championed by theoreticians and computer scientists. Many excellent books and monographs summarize these approaches [41–45]. However, the fact that simple systems can be modeled does not mean that by doing so we will be able to fully understand ourselves in the near future.

References

1. Aristotle. De Anima. Penguin Books; 1986.
2. Goodall J. Understanding Chimpanzees. 1st ed. Cambridge, MA: Harvard University Press; 1989.
3. Kandel ER. In Search of Memory. First ed. New York – London. W. Norton & Co.; 2006
4. Russell B. What is the Soul? In Praise of Idleness and Other Essays. 1st. ed. New York: Simon and Schuster; 1972:226–231.
5. Dietrich A, Been W. Memory and DNA. 71. J Theor Biol 2001;208(2):145–149
6. Miller CA, Gavin CF, White JA et al. Cortical DNA methylation maintains remote memory. Nat Neurosci 2010;13(6):664–666.
7. Steriade M. Coherent oscillations and short-term plasticity in corticothalamic networks. Trends in Neurosciences 1999;22(8):337–345.
8. Kandel ER. Cellular Mechanisms of Learning and the Biological Basis of Individuality. In: Kandel ER, Schwartz A, Jessell TM, editors. Principles of Neural Science. Fourth ed. New York: McGraw-Hill; 2000:1247–1279.
9. Byrne JH. Learning and Memory: Basic Mechanisms. In: Squire LR, Bloom FE, McConnell SK, Roberts JL, Spitzer NZ, Zigmond MJ, editors. Fundamental Neuroscience. Second ed. London: Academic Press; 2003:1276–1298.
10. Penfield W, Jasper H. Epilepsy and the Functional Anatomy of the Human Brain. Boston: Little, Brown and Co.; 1954.
11. Penfield W. The Excitable Cortex in Conscious Man. Springfield, IL: Charles C. Thomas; 1958.
12. Kandel ER, Kupfermann I, Iversen S. Learning and Memory. In: Kandel ER, Schwartz A, Jessell TM, editors. Principles of Neural Science. Fourth ed. New York: McGraw-Hill; 2000:1227–1246.
13. Eichenbaum HB. Learning and Memory: Brain Systems. In: Squire LR, Bloom FE, McConnell SK, Roberts JL, Spitzer NZ, Zigmond MJ, editors. Fundamental Neuroscience. Second ed. London: Academic Press; 2003:1299–1327.
14. Wilson MA, McNaughton BL. Reactivation of hippocampal ensemble memories during sleep. Science 1994;265:676–679.

15. Karni A, Tanne D, Rubenstein BS, Askenasy JJM, Sagi D. Dependence of REM sleep on overnight improvement of a perceptual skill. Science 1994;265:679–682.
16. Squire LR, Knowlton B, Musen G. The structure and organization of memory. Annual Review of Psychology 1993;44:453–495.
17. Panegyres PK. The contribution of the study of neurodegenerative disorders to the understanding of human memory. Qjm 2004;97(9):555–567.
18. LeDoux JE. Synaptic Self. How our brains become who we are. First ed. New York: Viking; 2002.
19. Fuster JM. Memory and planning. Two temporal perspectives of frontal lobe function. Advances in Neurology 1995;66:9–19.
20. Fuster JM. Distributed memory for both short and long term. Neurobiology of Learning & Memory 1998;70(1–2):268–274.
21. Knowlton BJ, Squire LR. Artificial grammar learning depends on implicit acquisition of both abstract and exemplar-specific information. Journal of Experimental Psychology: Learning, Memory, & Cognition 1996;22(1):169–181.
22. Cabeza R, Nyberg L. Neural bases of learning and memory: Functional neuroimaging evidence. Current Opinion in Neurology 2000;13(4):415–421.
23. Cabeza R, Nyberg L. Imaging cognition II: An empirical review of 275 PET and fMRI studies. Journal of Cognitive Neuroscience 2000;12(1):1–47.
24. Craik KJW. The Nature of Explanation. 2nd ed. Cambridge: Cambridge University Press; 1943.
25. Squire LR. Memory systems of the brain: A brief history and current perspective. Neurobiology of Learning & Memory 2004;82(3):171–177.
26. Reber PJ, Squire LR. Encapsulation of implicit and explicit memory in sequence learning. Journal of Cognitive Neuroscience 1998;10(2):248–263.
27. Schacter DL, Slotnick SD. The cognitive neuroscience of memory distortion. Neuron 2004;44(1):149–160.
28. Berkeley G. Philosophical Works. Rowman and Littlefield ed. Rowman and Littlefield; 1975.
29. Ryle G. The Concept of Mind. 1984 Reprinted ed. Chicago, Il: The University of Chicago Press; 1949.
30. Polanyi M. Personal Knowledge. Towards a Post-Critical Philosophy. Harper Torchbook ed. New York: Harper & Row; 1964.
31. Reber AS. Implicit Learning and Tacit Knowledge: An Essay on the Cognitive Unconscious. New York: Oxford University Press; 1993.
32. Knowlton BJ, Squire LR. The learning of categories: Parallel brain systems for item memory and category knowledge. Science 1993;262(5140):1747–1749.
33. Russell B. In Praise of Idleness. First ed. New York: Simon and Schuster; 1972.
34. Jacob F. Evolution and tinkering. Science 1977;196:1161–1166.
35. Jacob F. Complexity and tinkering. Annals of the New York Academy of Sciences 2001;929:71–73.
36. Bickle J. Psychoneural Reduction. The New Wave. 1st. ed. Cambridge, MA: The MIT Press; 1998
37. Kandel ER, Schwartz JH, Jessell TM. Principles of Neural Science. 4 ed. New York: McGraw-Hill; 2000.
38. Prigogine I, Stengers I. Order Out of Chaos. First ed. New York: Bantam Books; 1984.
39. Gell-Mann M. The Quark and the Jaguar. First ed. New York: W. H. Freeman and Company; 1994.
40. Papineau D. Mind the gap. In: Tomberling JE, editor. Philosophical Perspectives, 12, Language, Mind and Ontology. First ed. Malden, MA: Blackwell Publishers, Inc.; 1998:374–388.
41. Churchland PS, Sejnowski TJ. The Computational Brain. Cambridge, MA, MIT Press; 1992.

42. Abbott L, Sejnowski TJ. Introduction. In: Abbott L, Sejnowski TJ, editors. Neural Codes and Distributed Representations: Foundations of Neural Computation. 1st. ed. Cambridge, MA: MIT; 1999:vii–xxiii
43. Koch C, Davis JL. Large-Scale Neuronal Theories of the Brain. First ed. Cambridge, MA: The MIT Press; 1994.
44. Bickle J, Mandik P. The Philosophy of Neuroscience. The Stanford Encyclopedia of Philosophy, 2002. http://plato.stanford.edu/archives/win2002/entries/neuroscience/).
45. O'Reilly RC, Munakata Y. Computational Explanations in Cognitive Neuroscience. Understanding the Mind by Simulating the Brain. First ed. Cambridge, MA: The MIT Press; 2000
46. Kandel ER. Neuroscience – The molecular biology of memory storage: A dialogue between genes and synapses. Science 2001;294(5544):1030–1038.

8 Abstract and imaginary objects

Summary The classification of abstract and imaginary objects is a major source of contradictions among philosophers, who mistakenly believe that these objects are non-physical. Descartes thought that there were fundamental differences between *physical* and *mental* objects, and many of his contemporary philosophers also believed in *supernatural spiritual* objects. In contrast, the outstanding mathematician Gottlob Frege postulated that numbers belong to an *abstract* realm that has only *negative* properties. He acerbically criticized John Stuart Mill for confusing *pure* arithmetical propositions with the *practical use* of arithmetic. However, the written and archeological evidence from early arithmetic and geometry, and the developmental studies of Piaget, contradict Frege's ideas. In addition, neuroscience has demonstrated that abstract, imaginary, and metaphysical objects are the product of abstraction, which is physically realized in the frontal lobes of the brain.

8.1 The variety of objects

Philosophers have always been preoccupied by classifying *all that exists*, but opinions and classifications differ widely. Following Descartes, many have accepted that there is a fundamental difference between *material or physical* objects[1] (res extensa) and *mental* objects (res cogitans). Today, more than 90% of Americans believe in supernatural or *spiritual objects*, which are thought to include souls, God, and a variety of spirits, such as guardian angels and the devil. This implies that, in addition to physical and mental objects, there is an imaginary third realm of *spiritual or supernatural objects*.

---

[1] Some physical objects are not concrete, because they are forms of matter-energy that we cannot sense and can only detect using specific instruments, or by the indirect effects that they produce. Examples include subatomic particles, X-rays, ultraviolet radiation, rare gases, etc.

J.M. Musacchio, *Contradictions: Neuroscience and Religion*,
Springer Praxis Books, Popular Science, DOI 10.1007/978-3-642-27198-4_8
© Springer-Verlag Berlin Heidelberg 2012

As I will discuss below, the outstanding mathematician Gottlob Frege (1848–1925) made a different classification of objects that does not overlap with the previous one. He postulated that, in addition to the *mental* and physical objects, there is a realm of numbers, which are neither mental nor physical [1]. Philosophers who sympathize with rationalism believe that there are at least three different kinds of objects, *physical, mental, and abstract*, which do not exactly correspond to Frege's three realms. The acceptance of physical, mental, abstract, and supernatural realms seems an easy way to accommodate all extant objects, but it also creates some problems.

Catholics and members of other religions believe that humans have a soul, a supernatural substance that animates the body, will survive death, and will be judged by God to determine whether it will go to heaven or to hell. The idea that consciousness is a substance originates from what philosophers call reification. This consists in treating complex processes, whether natural or supernatural, as if these processes were manifestations of a single substance of unique nature.

The widespread use of the words "consciousness" and "soul" has most likely been influenced by religious ideas, because there is absolutely no objective evidence that souls or consciousness could be entities independent of the physical brain. From the neuroscience perspective, consciousness is neither a substance nor a simple object, but a dynamic collection of interrelated functions that take place only in animals with complex nervous systems (Chap. 5).

Mental and abstract objects include those objects physically realized in brains, such as our thoughts or experiences, including pains, desires, and smells. Even if these mental objects are initially perceived as being different from concrete physical objects, such as tables and chairs, neuroscientists have shown that they are nevertheless of a physical nature because physical agents can alter them. Additional subclasses of mental objects are the *imaginary* or *fictional* objects, because they seem to have been created out of nothing, as *hypothetical* objects. Imaginary or fictional objects include all objects of mythology and fiction, such as winged horses, goblins, unicorns, and cartoon or literary characters such as Mickey Mouse and Sherlock Holmes. Fictional objects are similar to abstract objects in the sense that they are not concrete and do not seem to exist *as such* in the objective world, even though some of their images are physically represented in paintings, tapestries, and movies, or in Disneyland and in a variety of toys. However, fictional objects are *not created out of nothing*. For example, Superman was created by combining in one character a variety of positive physical, mental, and moral properties, which were intellectually abstracted from different concrete cases.

Hypothetical objects are initially imaginary objects that play an important heuristic role in science. These objects are used to explain empirical anomalies that must logically have an identifiable cause. There are countless examples of hypothetical objects in science and medicine. We should remember that initially, toxins, germs, and hormones, were hypothetical objects which turned out to be real after further search. Hypothetical objects are initially postulated to explain the occurrence of certain anomalies and diseases. An outstanding example was provided by the German bacteriologist Robert Koch (1843–1910), who discovered the anthrax disease cycle in 1876, and the tuberculosis bacillus in 1882, for which he received the Nobel Prize in 1905. Another beautiful example of a hypothetical object, this one in astronomy, was postulated to account for the irregularities observed in the orbit of Uranus, which prompted the discovery of the planet Neptune, independently predicted by Le Verrier of France and by the British mathematician John C. Adams in 1846. Thus, historically, there have been innumerable hypothetical objects such as germs, planets, and genes which, thanks to their causal efficacy, were later proven to have a concrete physical existence.

The concept of *abstract objects* could also be ambiguous because in common language *abstract* means the opposite of concrete. Moreover, the word abstract fosters the belief that abstract, mental, or psychological objects are *nonphysical entities*. However, this is not the case because they have causal efficacy. Neuroscientists believe that *abstraction* is one of the most characteristically *human cognitive abilities*, indeed, the epitome of intelligence that allowed our species to build civilization, science, and culture. Abstraction is a complex physical brain function which serves to generalize and identify categories, concepts, and properties, which are in turn mental models we use for understanding and predicting reality. They are generally acceptable without much discussion and are conceived as mental snapshots that are useful because they seem to portray *the concrete* by fitting it into a higher, more general category of objects.[2]

In philosophy, the situation is more complex, because the nature of the abstract and its causal efficacy must be explained and justified. Some philosophers and mathematicians, of whom Frege was the most illustrious example, believed in a realm of *nonphysical* and *non-mental* abstract objects which could explain mathematics, logic, and ultimately everything there is. The implication of this assumption is that abstract objects consist in nonphysical entities with an independent status. However, this creates the problem of accounting for the nature of these entities and for their *causal efficacy* to interact with other realms, since nonphysical entities do not exist. Thus, the belief that mentally abstracted objects are actually abstract in the nonphysical sense is an illusion. What we call abstract objects, properties, or categories are the products of *abstraction*, a (physical) brain function that allows us to encode, organize, and then understand reality. There is a critical difference between referring to abstract objects as mentally abstracted, as opposed to *metaphysically* abstract objects. Metaphysically abstract objects are assumed to be members of an

---

[2] Frege thought that numbers were *abstract* in the metaphysical sense, and did not refer to them as abstract because they were mentally *abstracted* from the physical world.

independent realm of *abstracta*, which, because they lack causal efficacy, can neither be demonstrated nor interact with the physical realm. However, what we call metaphysically abstract objects are physically realized in brains.

The theories about the nature of abstract objects in mathematics and the multiple objections against metaphysically abstract objects will be discussed in the remainder of this chapter. It will be proposed that, contrary to Frege's belief, neither reason nor experience could detect the existence of "causally inert nonphysical objects". Abstract and imaginary objects, in the light of neuroscience and cognitive science, are objects *physically* encoded in the human brain. Thus, abstracted objects, like properties, must be physical to account for their mental efficacy and explanatory[3] power.

8.2 Frege's views of abstract and mathematical objects

During the twentieth century, the philosophy of mathematics was characterized by the belief in abstract entities, which included numbers, sets, propositions, and geometrical figures. Even though not all of today's mathematicians agree, the belief in *metaphysically abstract objects* is widespread. As indicated, their most ardent proposer and defender has been Gottlob Frege, who was passionate in his arguments against psychologism, a doctrine contending that psychological studies can provide the bases to logic, mathematics, and epistemology. Frege postulated that, in addition to the *realms of physical and mental objects*, there is a third independent realm of metaphysically *abstract objects*. He believed that there is an abstract realm composed of logical objects, truth-values, numbers, and other mathematical objects [2]. Frege thought that this realm was knowable through reason, which is related to logic, but according to him, certainly not to psychology. However, he never explained what he meant by reason. It seems that in addition to being full of contradictions, Frege was often inconsistent in his beliefs.

Frege postulated that *abstract objects* have a collection of *negative properties*, which include being non-spatial, non-temporal, non-mental, and nonphysical or non-detectable, and in consequence causally inert. Moreover, their existence was also presumed to be independent of any person thinking about them. Frege wrote, "Number is not abstracted from things in the way that colour, weight and hardness are, nor is it a property of things in the sense of what they are" [1]. He left unanswered the critical question of how objects with negative properties could interact with different realms or be recognized and used by

---

[3] Explaining–in common language—consists in providing the causal, logical, or psychological reasons for an occurrence or action, without getting involved in further analysis. Examples of abstract notions include beauty, the rights of minorities, the gross national product, the survival of the fittest, etc.

humans. It seems evident that some of Frege's thoughts were illogical and marred by contradictions.

To demonstrate that mathematics was a "superior science" (sic), Frege strived to minimize the influence of psychology and external aids, such as words or even numerals, in order to keep thoughts "more pure" . . . "[I]t is precisely in this respect that mathematics aspires to surpass all other sciences, even philosophy." He also stated that "psychology should not imagine that it can contribute anything whatever to the foundation of arithmetic".[4] Frege believed that relying on psychology makes "everything subjective, and if we follow it through to the end, does away with truth". Later, he states, "I found myself forced to enter into psychology, if only to repel its invasion of mathematics" [1].

Frege criticized John Stuart Mill for confusing *pure* arithmetical propositions with the *practical use* of arithmetical propositions. It seems that Frege believed that "pure" mathematics becomes "impure" when applied. He also said that the general laws of addition could not be laws of nature. Frege was deeply hostile when referring to the ideas of J.S. Mill on the origin of mathematics. He wrote:

> What, then, are we to say of those who, instead of advancing this work where it is not yet completed, despise it, and betake themselves to the nursery, or bury themselves in the remotest conceivable periods of human evolution, there to discover, like John Stuart Mill, some gingerbread or pebble arithmetic! It remains only to ascribe to the flavor of the bread some special meaning for the concept of number. [1]

It is clear that Frege was not naive and uninformed, but driven by his hostility to the idea that psychological abilities had anything to do with mathematics. When discussing the relevance of psychological experiences to mathematics, Frege completely abandoned his philosophical attitude and became dogmatic, sarcastic, and paranoid. The idea that numbers might have originated by abstraction to symbolize the numerosity of external objects was unacceptable to him because this would destroy the possibility of idealizing numbers into *metaphysically abstract objects* and would downgrade the presumed "superiority" of mathematics. Frege even objected to some of Kant's intuitionist views of arithmetical propositions. However, Frege excuses Kant because "Kant, obviously, was thinking only of small numbers." However, Frege realized that it is "awkward to make a fundamental distinction between small and large numbers." Frege was actually aware that Kant was right and that small numbers and simple arithmetic operations are evident to intuition, whereas this is not true for large numbers.[5] Frege's awareness is implied

---

[4] Note that, to stress his point, Frege anthropomorphized mathematics and psychology.

[5] Frege was mistaken in stating that large and small numbers were not intuitively different. A small number of objects (1 to 4–6) is immediately perceived by subitizing, so it is reasonable to believe that small numbers are encoded as analogue surrogates in our brains; however, larger numbers only exist encoded algorithmically as symbolic equivalents. As I will explain,

in the selection of the examples that he used to discredit J.S. Mill's approach and the intuitive concept of self-evidence. In his discussion, Frege asked . . . "is it really self evident that $135664 + 37863 = 173527$?" [1]. Obviously it is not, but as J.S. Mill could have said, it is evident that

$$\bullet + \bullet\bullet = \bullet\bullet\bullet \text{ and that } \bullet\bullet\bullet = \text{III or three}$$

These examples show that the unacceptability of psychological notions in the philosophy of mathematics can be traced to Frege's obsession against psychological explanations of mathematical objects and operations. The philosophy of mathematics might have progressed much faster if Frege had had his own children or taught arithmetic to children by counting fingers, pebbles, or gingerbread cookies. Curiously, despite his hostility toward psychology, Frege constantly relied on psychological or mental operations as explanatory; for example, he frequently mentions in his *Foundations of Arithmetic* the "eyes" of the mind, the "knowledge" of concepts, the meaning of words, intuition, self-evidence, reason, mental activity, logical faculty, etc. [1]. In disregarding psychology as a clue, Frege was forced to introduce *metaphysically abstract objects* without any justification, as a *Deus ex machina*. He just postulated the existence of an independent *metaphysically abstract realm* that was internally consistent, without realizing that a truly nonphysical realm should be *unknowable* because of its lack of causal efficacy.

8.3 Evidence against the metaphysically abstract nature of mathematics

Once we have the basic mathematical intuitions obtained from a small number of objects, we can extrapolate the rules to more complex arithmetical propositions and operations. A few numerals (symbolizing fingers or hands) and signs for simple operations are obviously sufficient to capture the essence of arithmetic, as shown in the objections to Frege's views; no ad hoc laws need to be invoked.[6] To understand why logic and mathematics have something to say about the world, it is necessary to mentally abstract arithmetic and logic from the physical world, à la Piaget [5–7]. Otherwise, there are no bases for explaining why nature is

---

we can estimate small numbers of objects accurately, without counting them explicitly. Actually, infants and even some animals, which do not have a language, can also subitize to a limited extent [3, 4].

[6] It is naive to believe that there are natural laws that must be "obeyed" by physical and abstract objects. In addition, it has never been demonstrated that natural laws are prescriptive, like the human laws or the Ten Commandments. The way in which energy transforms cannot be controlled by entities such as numbers and laws conceived as causally inert. We should consider natural laws only as descriptive of the way in which matter-energy transforms itself.

logical and describable through *mathematics*. Anthropology and early geometry also provide evidence against the metaphysically abstract nature of mathematics.

8.3.1 Anthropological evidence from early arithmetic

There is no doubt that the purpose of early arithmetic was to solve the accounting problems of everyday life. This view contrasts sharply with the view of the *metaphysically abstract* nature of mathematics. Studies of the early notation and the names of numbers point to the empirical roots of elementary arithmetic. The notation for numbers in early cultures clearly indicates that the symbols used represent the objects counted, in a one-to-one correspondence (Fig. 8.1). This is supported by many prehistoric archeological records that show that early humans recorded events and kept tally of objects by scoring sticks, rocks, and animal bones.

A 30,000 year old wolf bone found in Czechoslovakia has 55 notches in groups of five with a longer notch at 25. This has been interpreted as a rudimentary base five numbering system [8], indicating that the human hand was probably the initial model. Keeping tally of days was necessary to develop a calendar and to predict the seasons. The practice of making cuts or notches as a counting aid was probably widespread. A "score", as a term to denote a group of 20, is an archaic term, and presumably derives from the practice of counting sheep or large herds of cattle. The procedure consisted in counting orally up to 20 animals, and making a score or notch on a stick, before proceeding to count the next 20. The term "score" has today several meanings describing the accumulation of points obtained in sports or in any activity in which units are assigned to quantify performance (OED).

The earliest numerals were written as simple straight marks or lines, as is still evident in the first few Roman numerals (Fig. 8.1). Similar notations were used by the Babylonians, the early Egyptians, ca. 3400 B.C., and the Mayans. Since

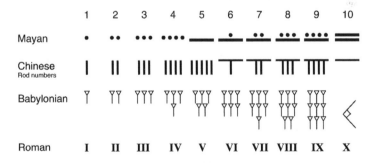

Fig. 8.1   Representation of numbers by early cultures. The numerals 1–3 or 1–4, even up to 10, were represented in a one-to-one fashion, using fingers and hands, as children do today, but as cultures evolved, they developed symbolic systems that culminated with the Arabic numerals and the zero that we use today

all the early notations for small numbers were *analogue one-to-one representations* of objects, *early numerals were representational*, and the marks were read by adding them. A special notation was also used in most systems to represent the numeral 10 and its multiples. The notation for ten, which *symbolizes* ten objects, implies the development of a rudimentary idea of multiplication. Moreover, it indicates that *symbols can also be used to symbolize symbols of higher order*. This conclusion seems trivial today, but the capacity to invent symbols and the additional step to symbolize symbols with *symbols of higher order*—the nesting of symbols—is perhaps one of the highest abilities of our brain.

The invention of the Arabic numerals 1–9 to substitute for marks or fingers that must be added represents a high level of abstraction that took several centuries to develop. The recurrence of the bases 10 or 20 in independent cultures indicates that the fingers, and in some cases the toes, were used for elementary calculations, as children do today. In addition, there are many linguistic indications that this was the case. For example, in several languages, most numbers between 10 and 20 are composed by *adding* a number to 10. In English, we add a number to -*teen*, the Old English suffix that stands for 10. This is also true in Spanish: *diecisiete* (17), from *diez* (10) and *siete* (7). In French, a remainder of a vigesimal (base 20) system is apparent in *quatre-vingts* (80). There is evidence that *fist* and *five* have similar roots, not only in English, but also in other languages, such as Indo-European, Dutch, and German. The words digit, digital, etc., from the Latin *digitus*, provide further indications of the origin of the early numbers, and leave little doubt that fingers and hands were used as the first practical calculator.

Some of the oldest mathematical developments were documented by the scribes in ancient Mesopotamia, who wrote on durable clay tablets [8]. There may have been older notations, but they have not survived. The Sumerian kingdoms in the third millennium B.C. developed mathematics for practical purposes, such as land allocation, business transactions, record keeping, tax accounting, etc. Their procedures were well established and standardized in the tablets, indicating that the initial steps of their development took place much earlier, but that they were not well documented. The Sumerians adopted first a decimal system, similar to that of the Egyptians, but the Old Babylonians, in the second millennium B.C., converted this system to a combination of decimal and sexagesimal (base 60). Since the Sumerians used a place-value notation, without a symbol for zero, they indicated its existence in calculations by leaving a gap [11].

Multiplication and division were extremely difficult without using the zero, so calculations were performed with the aid of abbreviated pre-calculated *tablets*. This probably accounts for the current expression multiplication *tables*. The use of the base 60 facilitated division and multiplication (by 2, 3, 4, 5, 6, 10, 12, 15, 20, and 30), and allowed the solution of much more complex arithmetic problems. The sexagesimal system became important for the development of astronomy between the third to first centuries B.C. [8]. The system proved to be so useful that its units of *minutes* and *seconds* are still used for measurements of *time* and *angles*. Interestingly, Babylonian mathematics was made up of sets of rules to manipulate data, and did not invoke theorems to demonstrate the validity of the methods used. Babylonian mathematics remained empirical, but it was for the most part accurate and reliable. Moreover, the treatment of algebraic equations in Babylon was highly elaborate, but restricted to individual problems, without evidence for a general procedure. Their algebraic abilities were never matched by the Greeks, even in their Golden Age [9, 10].

The most remarkable aspect of the Greeks, in contrast to the Babylonians and Egyptians, is that they were more interested in theoretical rather than practical applications of mathematics. Their efforts concentrated on geometry, but it eventually permeated their arithmetic. The Greek numerals were alphabetic, and they used letters to represent numbers, supplemented by a few symbols borrowed from the Semitic. However, the Greeks may have received more credit than they deserve because they were not very generous in recognizing the contributions of their predecessors.

The Romans developed a numerical system in which letters of the alphabet were used to represent numbers up to 100,000. The modern version is the familiar one: I = 1; V = 5; X = 10; L = 50; C = 100; D = 500; M = 1,000. They developed a rudimentary place-value notation in which the numbers were added to calculate the final value, but when a small number preceded a large one, it was subtracted from the following larger symbol. Early cultures operated only with small numbers, but as their needs grew, they developed ways to express larger numbers. The largest Roman number that we use today is M, but in the Roman era, a horizontal line added over the set multiplied it by 1,000; vertical lines on both sides multiplied it by 100 [8].

In the Mayan vigesimal system (base 20), the first four numerals were written as dots, again in a one-to-one representation. Five, the equivalent to one hand, was represented by a single horizontal line, indicating that the Mayan system attained a high level of abstraction independently of other cultures (Fig. 8.1). One horizontal line is repeated for every five units, until three lines and four units are accumulated in a *vertical* fashion. This notation corresponds to 19, the highest number that could be written in the first place-value. This example illustrates once again that the symbols written within a certain space were *added* to read the final value. The Mayans also independently invented zero, but apparently they did not use it for advanced computations. Numbers in the second place-value were multiplied by 20 as opposed to 10 in our decimal system. The Mayan number system was used mainly to calculate the calendar, so the third place-value was not the product of 20×20 (or 20 [2])—as in a regular base 20 system—but 18×20, which is 360, the approximate number of days in a year. The Mayan calendar consisted of 18 months of 20 days each, and five additional days that were added to complete the year. The different place-values of the Mayan numbers were written vertically.

Calculations with all the early numbering systems were very difficult, so several versions of the abacus were developed. The source of our word *abacus*, the Greek word *abax*, probably comes from the Hebrew word for "dust." One of the meanings of the Greek word is "a board sprinkled with sand or dust for drawing geometric diagrams," which can be used as an aid in discussions (OED). The Romans used a version of the abacus consisting of a *counting* board with grooves, in which the numbers were represented by little stones or *calculi*, the origins of the roots for the words "counter," "calculation," etc. The Roman counting board is a precursor of the abacus with which we are now so familiar, which has movable counters strung on rods. Different models of *abaci* were used by the ancient Egyptians, Greeks, and Romans, as well as by the Orientals. The abacus was used until several decades ago by the older practitioners in small shops, but they were rapidly replaced by electronic calculators and later by computers.

The place-value notation—the change in the value of the symbols indicated by a change in relative position (units, tenths, hundreds, etc.)—is a remarkable intellectual achievement, which indicates a higher level of abstraction than the one-to-one representation by scratches, pebbles, or the numerals from 1 to 10. The place-value notation is a symbolic device that was developed independently by several cultures. As indicated, it was used by the Mayans, who also independently invented zero, although they did not use it for calculations, but rather for their calendar. In India, the use of zero can be traced back to the third century B.C., but it was not until the sixth century that the use of the nine Arabic numerals and zero became more general [11]. This system facilitated calculations enormously, by using simple algorithms and calculating techniques that gradually became universal.

The early numeral notations show that, contrary to Frege's ideas, numbers were invented to count and organize objects in a one-to-one relationship. Numbers are symbols that replace marks, cuts, or notches, originally represented by fingers, which in turn were used to count, and probably later to organize objects in a sequence. Thus, it is crucial to understand the complexity and the nesting of higher order symbols, to see that *the one-to-one relationship with objects is symbolically maintained*, even if we use Arabic numerals. The reliability and universality of basic arithmetic is amazing, probably because initially its results were constantly checked and rechecked, and reliable calculations were essential for trading. As Owen says in his magnificent book [11], "The experience of ages is built into them, and the repetitive usefulness of the operation of addition provides a uniform aspect to all". The basic arithmetical operations that model physical process were verified empirically in millions of experiments long before mathematicians tried to put arithmetic on a "solid" basis. It is interesting that comparable solutions to similar problems were found in independent cultures. This is reminiscent of the *convergent evolution* observed in genetically different animal species and which results in the formation and evolution of similar organs when they have been long exposed to similar environments.

As indicated above, numerals were initially symbols that represented quantity or order of objects. Later, mathematicians defined several different kinds of numbers, some of which are derived from constant relations, such as π, e, square roots, etc. Numbers can also represent relations of higher order "relations of relations", etc. However, numbers are only symbols for something else, including relations of different complexity. Numbers are products of our brain that were independently created by different cultures and later fused in our universal mathematics. The power of numbers, however, fascinated all early cultures, and Pythagoras is the most famous of the earliest mathematicians who attributed some supernatural properties to numbers [8, 12].

8.3.2 The evidence from early geometry

The Greeks learned geometry (*geometrein:* to measure land) from the Egyptians, who developed elaborate surveying techniques to cope with the blurring of property boundaries caused by the periodic floods of the river Nile. The Egyptians in turn learned some of their methods from the Mesopotamians, who had similar problems caused by the floods of the Tigris and the Euphrates. Egyptian surveyors were also known as the "rope-stretchers," because their main instrument of measurement was a rope with knots at equally spaced intervals. Surveyors in Mesopotamia, like those in Egypt, used the "Pythagorean relation" (the 3, 4, 5 right-angled triangle) to lay out right angles, long before the time of Pythagoras (Fig. 8.2).

Despite its use, there are no indications of a general demonstration of the theorem. The Egyptians were not interested in the abstract aspects of mathematics. The Greeks recognized their debt to the Egyptians, and developed the theoretical bases of geometry. Translations of Babylonian tablets also suggest that ..."the Greeks built within the scaffolding of early Babylonian algebra and number theory" [11]. The "idealization" of mathematics started in the times of Pythagoras, who was credited with inventing and stressing the importance of *mathematical proofs* in the sixth century B.C. He also realized that, in order to deduce a proof, it was first necessary to lay down postulates and assumptions [11], which were undoubtedly empirically derived. These ideas had a profound effect on the future of mathematics, establishing the methods and the models that mathematicians followed for many generations.

Euclid of Alexandria, who lived in the third century B.C., compiled and systematized the knowledge of geometry and arithmetic that had accumulated over at least two centuries before his time [13]. Euclid adopted five postulates,

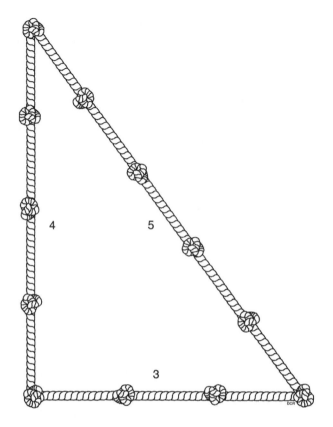

Fig. 8.2 The Rope Stretchers. Representation of a rope triangle with sides of lengths 3, 4, and 5, which was used by the "rope-stretchers" to lay out right angles. Such triangles were necessary to orient buildings and to retrace the limits of farms after the periodic floods of the Nile in Egypt, or the Tigris and Euphrates in Mesopotamia. The Pythagorean relationship was used by the rope stretchers long before Pythagoras demonstrated his theorem

considered self evident or acceptable without proof. The postulates, together with five additional rules, the common notions or axioms, were used to demonstrate theorems, from the simplest to the most complex. In this fashion, all classic geometry was "deduced" from the initial assumptions. Euclidean geometry has been considered one of the most remarkable feats of the human mind, the highest example of the deductive sciences, which has served to educate countless generations of children.

The body of work providing the bases for Euclidean geometry is certainly impressive, but the beautiful construction had some structural weaknesses which had become evident by the end of the nineteenth century. Euclid's system was deductive, but it was based on initial intuitions which required a series of definitions, axioms, postulates, and common notions as starting point.

Euclid's "definitions" describe the constitutive elements of geometry, that is, the already known lines, angles, polygons, etc. Lines, points, and angles are all geometrical elements that were quite evident in the external world and in the work of the rope-stretchers. Of course, real world structures are never perfect, and real lines have a certain breath or body, but they are easily imagined as pure forms, ideal objects, and independent of any physical reality. Thus, it is not surprising that Euclid was able to create abstract models of the physical two-dimensional space, as it is empirically perceived.

Euclid and his school defined the elements of geometry in such a way that these elements were idealized to the point that they lost all their material attributes and became pure forms. From a line traced in the sand, they produced a pure abstracted form defined as a "breadthless length," or "length without width." Euclid's "definitions" were actually *abstractions* or idealizations of geometric elements that existed in the external world. They were consistent with Plato's conception of things, according to which experiences help to remember truths that were already present in the soul from a previous, purer existence. Davis [10] has indicated that, besides the definitions and the axioms, an axiomatic system requires *undefined terms* to avoid infinite regress, because all closed systems of definitions are circular. Thus, geometry is also based on intuitions so it is as *incomplete* as arithmetic or language, because it ultimately relies on experiences (see Chap. 6). Euclid missed this point because his definitions were idealizations of objects known to everybody by the time they encountered his geometry. Thus, in Euclid's system, experiences and implicit knowledge stood in for undefined terms.

Euclid's *axioms* or *postulates* were undoubtedly the most impressive set of abstractions produced in the third century B.C. The postulates were probably quite reasonable to his students and to the contemporary surveyors, who accepted them because they were abstractions of elements from a concrete and familiar reality. Moreover, the postulates were *empirically selected as the minimal number necessary to produce the desired demonstrations*. The list of axioms and postulates is not identical in all editions of the *Elements*, because postulates and axioms were shifted from one list to another by different editors. The *common notions* of Euclid—or axioms as they are called by some—also vary in the different translations of *Euclid's Elements*, according to Sir T.L. Heath, who wrote an introduction to Todhunter's edition of Euclid [13]. The original five common notions attributed to Euclid are listed in some editions as up to ten, with the additional ones taken from the postulates.

The capacity to generalize or extrapolate is the key component of the geometric method used by Euclid. The first approximations, however, must be derived empirically, from the idea and construction of the elementary geometric figures to the division of the right angle into 90°. There is no way to have Euclidean geometry without the *intuitions generated* by empirical knowledge. The initial intuitions and the invention and conception of ideal forms, which were derived from the imperfect and irregular figures constructed in the external world, were essential steps in the overall development of geometry. Euclid's work has been instrumental in establishing mathematical methodology and in setting the stage for the development of "abstract" mathematics. The method of Euclid had a remarkable heuristic value in laying the foundations for further developments of the axiomatic method. By changing one of the "axioms", and still using the deductive method, Lobachevsky and Bolyai, working independently, invented the non-Euclidean geometries [8]. However, Euclid's geometry, despite all its empirical roots and faults, can be construed as the mother of all geometries.

Once the symbolic system has been constructed, the definition of some of the objects, or the axioms abstracted from the system, can be changed, and new geometries can be derived. Regardless of which system a new geometry fits, *geometry* is the last word in the name of all the non-Euclidean geometries, a testimony to their empirical origins. The existence of the symbolic structures of arithmetic and geometry is certainly not a proof of their *abstract* nature. They are empirical abstractions from the physical world, which were perceived by the senses, and physically encoded in our brains.

8.4 Intuitions and abstractions transport us to a symbolic universe

The capacity of making abstractions is perhaps the most characteristic and evolved of our abilities. It permeates practically everything that we do from making trivial decisions to projecting ourselves into the future by planning our long-term goals. Intuitions and abstractions are essential to identify the important properties of complex subjects and to be aware of moral and esthetic values. Moreover, abstraction is what partially liberates us from the tyranny of our concrete organic needs. We entered the world of culture and values only through our capacity to transcend and see beyond the concrete reality, and to imagine and construct a world of symbols and values represented by the culture in which we live. As Ernst Cassirer said, we live in our own cultural and symbolic universe; we should define ourselves as *symbolizing* animals, rather than as rational animals [14].[7]

Our knowledge of symbols and our capacity to encode and transmit them makes it possible to transfer some of our experiences and discoveries to our descendants. Thus, cultural inheritance through different symbolic forms becomes the most important addition to our genetic endowment. The cultural endowment relies on our cognitive ability to make abstractions and grasp the essential characteristics of complex cultural, social, and economic situations.

---

[7] Interestingly, Ernst Cassirer was already aware of the work of Kurt Goldstein (see below) when he wrote "An Essay on Man" [14].

This abstract ability is uniquely human and is the product of our highly developed prefrontal lobes, as established by several neurologists, from Kurt Goldstein to Joaquín M. Fuster [15–17].

Children learn about metaphysically abstract objects indirectly, by first learning through intuition and generalization[8] from concrete examples taken from arithmetic, geometry, and logic. The importance of intuition in apprehending abstract notions has been recognized by many authors, including Kant as indicated by Cassirer [14] and Mill [18]. Even if Frege could not deny the overwhelming evidence regarding the role of intuition in geometry, he denied its equally obvious role in arithmetic [1].

The development of the concept of number and mathematical operations is a complex process that has been the subject of countless psychological studies during the twentieth century. These studies and the cognitive disorders produced by brain diseases help us to understand the acquisition of mathematical knowledge. Children develop the concept of number and arithmetical rules through what Piaget called reflective abstraction [19].[9] The process requires the active participation of the child in tasks such as classifying, ordering, and putting objects into correspondence. Piaget prefers the term "invention" to "discovery" of numbers, because the concepts are constructed by each child from within [7]. It is clear that sensory information and reasoning are necessary to learn to count and to understand simple arithmetical operations. One of the important landmarks in the *perception of quantity* by a child is the development of the idea of the *conservation of objects*, which is needed in order to develop the related idea of the *conservation of number*. Before this development, children tend to judge quantity by the space taken up by counters or beads. Thus, changes in the length of a row occupied by a fixed number of counters are initially misinterpreted as changes in quantity. The longest row of counters is interpreted as having more counters [20].[10]

Before children acquire the notion of number and arithmetical operations, they have an intuitive sense of one-to-one correspondence. Piaget calls intuitive any numerical correspondence that is entirely based on perception or on representative images [6]. He also called "perceptual numbers" those that can

---

[8] To understand biology and other natural processes we need first to describe them as objectively as possible. The acquisition of knowledge is the result of natural processes, so a priori normative conceptions cannot dictate how things are. Cognition is a natural process that phylogenetically and ontogenetically precedes the development of philosophy.

[9] The concept of reflective abstraction is extensively discussed in Kitchener [5]. Even though many of Piaget's conceptions are not acceptable today, his pioneering methodology has important implications for understanding the epistemology of mathematics and logic. See also Ref. [7].

[10] Interestingly, the rows must have more than seven counters to confuse the child. Small numbers, up to four or five, can be easily distinguished at a glance; they were called perceptual numbers by Piaget. Today, distinguishing numerosity at a glance is called subitization (Sect. 8.5 and [7]).

be distinguished at a glance, up to four or five objects [7]. Today, this language-independent process is called subitizing, and it is found not only in children [21], but also in several animal species.[11] This preverbal counting system provides the framework for understanding the verbal system [24]. Thus, language provides only a symbolic system to express a language-independent representation of perceptual numbers which we share with other species [25]. This does not negate the fact that language and written symbols can in turn contribute to the extrapolation and unlimited expansion of the original non-verbal concepts. Perceiving numerosity—subitizing—provides the empirical roots of arithmetic.

8.5 Subitizing: the implicit perception of numerosity

During the second half of the twentieth century, cognitive psychologists demonstrated that human infants and some animals have innate language-independent abilities to perceive numerosity [26–28]. This capacity has been referred to as the "number sense" by Dehaene [29], but it is a complex function composed of at least two processes. The first and most elemental process is *subitizing*, which consists in the accurate nonverbal perception of up to four or five objects [30].[12] The second is an analogue process for representing large, but approximate numerical quantities [27]. Subitizing and the analogue estimation of numerosity are two qualitatively different and separable processes, but there is disagreement as to whether or not there is a statistical break or a clear discontinuity between them [31].

Subitizing is thought to be an elemental symbolic process that makes counting possible by providing the bases for the elemental representations of numerosity. Together with the analogue estimation of numerosity, they may provide the bases on which numerical intuitions develop [32]. Subitizing is a fast, accurate, and language-independent process, whereas verbal counting is slower. In infants and animals, subitizing is unquestionably a non-verbal process, whereas counting of multiple objects by adult humans has an undeniable verbal component. Overall, these studies have established the innate non-verbal capacity to represent numerosity.

Studies in monkeys and humans indicate the involvement of at least two different cerebral networks that underlie the number sense [33, 34]. The existence of the two systems has been confirmed in humans with brain lesions.

---

[11] They include pigeons, parrots, rats, dolphins, monkeys, and chimpanzees [22, 23].

[12] The term was suggested by Dr. Cornelia C. Coulter and introduced by E.L. Kaufman. See the OED [26] and [27]. Subitizing was originally thought to determine up to seven or "seven plus or minus two" objects. However, recent studies have limited subitizing to the perception of four or five objects [30].

Patients with lesions in the parietal lobe (see Fig. 4.2) may have deficits in the basic functions involved in preverbal quantity representation, subitizing, and approximate addition and subtraction, as well as comparisons, but simple multiplication is relatively well preserved. In contrast, patients with aphasia and semantic dementia due to left fronto-temporal lesions near the area of Broca (Fig. 4.2) have severe deficits in multiplication, because of the link between language and memorized multiplication tables. Patients with aphasia may also have intact approximation abilities and processing of non-symbolic numerosities [35]. Cognitive scientists agree on the naturalistic approach to the foundation of mathematics, even though they disagree on how the representations take place and how they are organized.[13]

Piaget was without doubt the first systematic observer of how children learn to understand abstract objects. He was probably the most influential person to champion the notion of *genetic epistemology*,[14] which has never been seriously considered by most professional philosophers. Actually, Piaget thought of himself as an epistemologist [5]. However, he made an important contribution to "naturalizing" epistemology by considering cognition a natural phenomenon that should be studied empirically. Despite some recent criticism, Piaget's theories and the stages of cognitive development that he described, have far-reaching philosophical implications. The crucial point is that, as Piaget proposed, logico-mathematical notions are derived from concrete experiences followed by abstraction. Their development is implicit and subtle, so if these processes are not specifically investigated, their empirical acquisition is overlooked. This is why some philosophers and mathematicians still believe that mathematics and logic are *a priori* sciences.

To define triangles, the classic procedure consists in idealizing all the material components of concrete triangles. Let us say that we are looking at triangles such as those drawn by Euclid on the sand, by schoolchildren on a paper, or as formed by a rope (Fig. 8.2). We can picture the ideal lines that determine the sides of the triangle by "mentally eliminating" all the physical traces left in the sand or in the paper. We can also conceptually generalize to triangles with different types of angles. Thus, we can think of "abstract triangles" as if they were "pure forms" without any physical element, as idealized by Euclidean geometry. We are left with a definition that includes an imaginary plane and three points that are connected by three lines that enclose three angles. Plane, points, lines, and angles are the products of previous abstractions, so we feel that we have succeeded in defining abstract, nonphysical triangles. Disembodied forms, like the abstracted Euclidean triangles, do not have *physical* existence. However, as discussed

---

[13] There are many papers on this subject: see Refs. [21, 25, 29, 32, 36–38].

[14] According to James Russell, the term genetic epistemology (which refers to genesis) was coined by James Mark Baldwin. Russell adds that several thinkers proposed similar ideas independently, all of which arose one way or another from Darwin's evolutionary theory [39].

previously (Chaps. 4, 7), memories and ideas are physically encoded in the brain; triangles cannot be considered metaphysically abstract objects, precisely because they are physically encoded in our brains.

The nature of numbers and geometric elements is not an academic curiosity, but instead has many implications that help to clarify the empirical bases of mathematics. Contrary to the preconceived ideas of some mathematicians and philosophers, Piaget's observations show that *logico-mathematical notions develop implicitly through concrete experiences*. However, not all the steps of these complex processes are known, and there are several theories about how we acquire arithmetical and geometric notions. Mathematical objects are symbolic objects created by abstraction, and referred to through common words and written symbols. A multidisciplinary approach shows that numbers and geometric objects are not derived from an independent *metaphysically abstract* realm, as Frege and others proposed.

8.6 The neural bases of abstraction

Abstraction is a neurobiological process that is essential to generalize and understand complex problems, but as such it must be studied empirically. Problems arise in relying exclusively on methods that ignore not only the empirical evidence, but also the physical nature of thinking. Initial clues regarding the role of abstraction in normal behavior were found by observing the effects of diseases on humans. As Kurt Goldstein indicated, "Nobody will doubt that the observation and analysis of pathological phenomena often yield greater insight into the processes of the organism than that of the normal" [16]. He added that, since diseases involve a profound modification of normal processes, we could not afford to ignore them. There is no doubt that diseases count among the most cruel experiments on humans, as described in Chap. 4. This should prompt us to study all the available evidence from humans, because animal experimentation cannot produce much insight into mental functions that are rudimentary or absent in lower animals.

Kurt Goldstein studied brain injuries that occurred during the First World War [15]. His patients were otherwise healthy, because their wounds had healed, so he was able to follow them for up to 8 years. Goldstein put forward the idea that patients with severe lesions of the frontal lobes lose their "abstract attitude" or the capacity to deal with abstract thoughts, and show an "impairment of the capacity to comprehend the essential features of an event." The English neurologist Henry Head also described similar changes as a disturbance of the patients' "symbolic expression". In addition, frontal lobe patients show deep changes in personality, similar to those of Phineas Gage, whose symptoms were described more than a century ago and were reinterpreted in the light of recent brain imaging studies and cognitive tests [40].

In contrast to the precise (somatotopic) localization of motor and sensory functions in the brain cortex, the symptoms of patients with lesions in the prefrontal areas are associated with a broad range of cognitive functions, which include problem solving, planning, and managing work and personal life, as well as judging complex situations. These cognitive functions have been globally described as "goal oriented behavior"[15] or "executive functions". Brain imaging studies indicate that the prefrontal lobes (Fig. 4.2) are involved in such functions, but there is no general agreement on the details [45]. The symptoms shown by frontal lobe patients are variable and even restricted lesions produce deficits in several different cognitive tasks which are difficult to separate from "general intelligence" [46]. This is attributed to the interaction of anatomically and functionally distinct systems in the performance of complex tasks [47]. In addition, the ventromedial areas of the prefrontal cortex affect decision-making through an emotional component [48]. Besides traumatic brain injuries, several conditions, such as normal aging, Alzheimer's disease, frontotemporal dementia, Parkinson's disease, strokes, and frontal lobe operations, have been described to interfere with some aspects of the executive functions.[16] The abundance of clinical data, which has been repeatedly verified in neuropathological, surgical, and brain image studies indicates that abstraction is a function that is implemented by complex neurophysiological processes. Our abstract capacity is the product of physical brain functions that can transport us to the most sublime symbolic universes.

8.7 Summary and conclusions

The nature of abstract and imaginary objects is potentially confusing, because the imperceptibility of the neural mechanisms that implement abstraction creates the illusion that abstract objects encoded in the brain are non-physical objects. Descartes and other rationalist philosophers, near the end of the nineteenth century, distinguished the existence of physical and mental objects. The mathematician Gottlob Frege lived in an imaginary world full of contradictions and composed of physical and mental objects, with an additional third realm composed of abstract objects, which included numbers. According to Frege, numbers were not abstracted from the external world but

---

[15] Several anatomic circuits that link the frontal cortex to the striatum, globus pallidus, substantia nigra, and thalamus are able to produce deficits in the executive function [41, 42]. This is consistent with the report that, during category learning by abstraction, significant increases in brain activity have been found by fMRI studies in the left and right anterior frontal cortex and in the right inferior lateral frontal cortex [43]. In addition, some studies also describe activity in posterior brain regions, indicating the involvement of dynamic and distributed networks [44].

[16] These findings are consistent with the complexity of the distributed networks that subserve executive functions [49–54].

found by reason. However, the anthropological record and the historical evidence, including early arithmetic and geometry, show that numbers and abstract objects are mentally abstracted from empirical and concrete situations. The evidence from neurology and neuroscience also shows that the capacity of making abstractions and mental calculations are physical processes. Abstract objects, including numbers are the product of a characteristically human ability for abstraction that is mainly associated with neural circuitry in the prefrontal lobes. Abstract and symbolic objects retain their physicality in our brains, so they can transform the world, because they have causal efficacy. The evidence discussed solves the problem of understanding the power and nature of mathematics and eliminates numbers and abstract objects as members of a hypothetical immaterial realm. It supports the idea that matter-energy is the only existent realm.

References

1. Frege G. The Foundations of Arithmetic. A logico-mathematical enquiry into the concept of number. 2nd ed. Oxford: Basil Blackwell; 1974.
2. Burge T. Frege on knowing the third realm. Mind 1992;101:633–650.
3. Starkey P, Cooper RG. The development of subitizing in young children. British Journal of Developmental Psychology 1995;13:399–420.
4. Murofushi K. Numerical matching behavior by a chimpanzee (*Pan troglodytes*): Subitizing and analogue magnitude estimation. Japanese Psychological Research 1997;39 (3):140–153.
5. Kitchener RF. Piaget's Theory of Knowledge. Genetic Epistemology & Scientific Reason. First ed. New Haven: Yale University Press; 1986.
6. Piaget J. The Child's Conception of Number. New York: W. W. Norton & Company, Inc. 1965.
7. Kamii CK, DeClark G. Young children reinvent arithmetic. New York: Teachers College, Columbia University; 1985.
8. Bunt LNH, Jones PS, Bedient JD. The historical roots of elementary mathematics. Reprint ed. New York: Dover Publications, Inc.; 1988.
9. Bell ET. Men of Mathematics. Reprinted ed. New York: Simon and Schuster; 1937.
10. Davis DM. The Nature and Power of Mathematics. First ed. Princeton, NJ: Princeton University Press; 1993.
11. Owen GE. The Universe of the Mind. Baltimore: The Johns Hopkins Press; 1971.
12. Ore O. Number theory and Its history. 1st. ed. New York: Dover Publications, Inc.; 1988.
13. Euclid. Euclid's Elements. Books I-VI, XI and XII. Reprint ed. London: J.M. Dent & Sons Ltd.; 1961.
14. Cassirer E. An Essay on Man. An introduction to a philosophy of human culture. 1st. ed. New Haven and London: Yale University Press; 1944.
15. Goldstein K. Human Nature in the Light of Psychopathology. First ed. Cambridge, MA: Harvard University Press; 1951.
16. Goldstein K. The Organism. A holistic approach to biology. New York: American Book Company; 1939.
17. Fuster JM. The Prefrontal Cortex. Fourth ed. London, UK: Academic Press; 2011.
18. Mill JS. A System of Logic. Eighth, reprinted ed. London: Longmans, Green and Co.; 1911.
19. Piaget J. Genetic Epistemology. First ed. New York: Columbia University Press; 1970.
20. Piaget J. Psychology and Epistemology. First ed. New York: Grossman Publishers; 1971.

21. Carey S. Knowledge of number: Its evolution and ontogeny. Science 1998;282 (5389):641–642.
22. Brannon EM, Terrace HS. Ordering of the numerosities 1 to 9 by monkeys. Science 1998;282(5389):746–749.
23. Brannon EM, Terrace HS. Representation of the numerosities 1–9 by rhesus macaques (*Macaca mulatta*). J Exp Psychol Anim Behav Process 2000;26(1):31–49.
24. Gallistel CR, Gelman R. Preverbal and verbal counting and computation. Cognition 1992;44(1–2):43–74.
25. Gelman R, Gallistel CR. Language and the origin of numerical concepts. Science 2004;306 (5695):441–443.
26. Guttman R. Beyond subitizing—Is estimation of numerousness a continuous process. Acta Psychologica 1978;42(4):293–301.
27. Dehaene S. Varieties of numerical abilities. Cognition 1992;44(1–2):1–42.
28. Uller C, Hauser M, Carey S. Spontaneous representation of number in cotton-top tamarins (*Saguinus oedipus*). J Comp Psychol 2001;115(3):248–257.
29. Dehaene S, Dehaene-Lambertz G, Cohen L. Abstract representations of numbers in the animal and human brain. Trends in Neurosciences 1998;21(8):355–361.
30. Trick LM, Pylyshyn ZW. What enumeration studies can show us about spatial attention—evidence for limited capacity preattentive processing. Journal of Experimental Psychology-Human Perception and Performance 1993;19(2):331–351.
31. Balakrishnan JD, Ashby FG. Is subitizing a unique numerical ability? Perception & Psychophysics 1991;50(6):555–564.
32. Feigenson L, Dehaene S, Spelke E. Core systems of number. Trends in Cognitive Sciences 2004;8(7):307–314.
33. Dehaene S. Neuroscience. Single-neuron arithmetic. Science 2002;297(5587):1652–1653.
34. Nieder A, Miller EK. A parieto-frontal network for visual numerical information in the monkey. Proceedings of the National Academy of Sciences of the United States of America 2004;101(19):7457–7462.
35. Lemer C, Dehaene S, Spelke E, Cohen L. Approximate quantities and exact number words: Dissociable systems. Neuropsychologia 2003;41(14):1942–1958.
36. Wynn K. Psychological foundations of number: Numerical competence in human infants. Trends in Cognitive Sciences 1998;2(8):296–303.
37. Carey S. Cognitive foundations of arithmetic: Evolution and ontogeny. Mind & Language 2001;16(1):37–55.
38. Barth H, La Mont K, Lipton J, Dehaene S, Kanwisher N, Spelke E. Non-symbolic arithmetic in adults and young children. Cognition 2005.
39. Russell J. The Acquisition of Knowledge. 1st. ed. New York: St. Martin's Press; 1978.
40. Damasio H, Grabowski T, Frank R, Galaburda AM, Damasio AR. The return of Phineas Gage: Clues about the brain from the skull of a famous patient. Science 1994;264 (5162):1102–1105.
41. Mega MS, Cummings JL. Frontal-subcortical circuits and neuropsychiatric disorders. Journal of Neuropsychiatry & Clinical Neurosciences 1994;6(4):358–370.
42. Owen AM. Cognitive dysfunction in Parkinson's disease: The role of frontostriatal circuitry. Neuroscientist 2004;10(6):525–537.
43. Reber PJ, Stark CE, Squire LR. Cortical areas supporting category learning identified using functional MRI. Proceedings of the National Academy of Sciences of the United States of America 1998;95(2):747–750.
44. Carpenter PA, Just MA, Reichle ED. Working memory and executive function: evidence from neuroimaging. Current Opinion in Neurobiology 2000;10(2):195–199.
45. Duncan J, Owen AM. Common regions of the human frontal lobe recruited by diverse cognitive demands. Trends in Neurosciences 2000;23(10):475–483.
46. Duncan J. Frontal lobe function and general intelligence: Why it matters. Cortex 2005;41 (2):215–217.

47. Adcock RA, Constable RT, Gore JC, Goldman-Rakic PS. Functional neuroanatomy of executive processes involved in dual-task performance. Proceedings of the National Academy of Sciences of the United States of America 2000;97(7):3567–3572.
48. Bechara A. The role of emotion in decision-making: Evidence from neurological patients with orbitofrontal damage. Brain & Cognition 2004;55(1):30–40.
49. Bherer L, Belleville S, Hudon C. Executive function deficits in normal aging, Alzheimer's disease, and frontotemporal dementia. Psychologie et Neuropsychiatrie du Vieillissement 2004;2(3):181–189.
50. Buckner RL. Memory and executive function in aging and AD: Multiple factors that cause decline and reserve factors that compensate. Neuron 2004;44(1):195–208.
51. Duke LM, Kaszniak AW. Executive control functions in degenerative dementias: A comparative review. Neuropsychology Review 2000;10(2):75–99.
52. Lamar M, Swenson R, Kaplan E, Libon DJ. Characterizing alterations in executive functioning across distinct subtypes of cortical and subcortical dementia. Clinical Neuropsychologist 2004;18(1):22–31.
53. Lewis SJ, Dove A, Robbins TW, Barker RA, Owen AM. Striatal contributions to working memory: A functional magnetic resonance imaging study in humans. European Journal of Neuroscience 2004;19(3):755–760.
54. Miotto EC, Morris RG. Virtual planning in patients with frontal lobe lesions. Cortex 1998;34(5):639–657.

9 Nature is logical, because logic is natural

> ... [T]he universe is known to man only through logic and mathematics, the
> product of his mind, but he can only know how he constructed mathematics
> and logic by studying himself psychologically and biologically ...
> —J. Piaget, Psychology and Epistemology

Summary Nature is logical because elemental logic has been implicitly abstracted from natural processes. The symbolic representation of logical statements shows a "form" that logicians have long believed to guarantee the truth of statements. This is most applicable to *deductive* logic in which the conclusions are implicit in the premise. In contrast, *inductive* logic cannot be formalized and is inferred from many specific observations to make a general hypothesis, whose validity can be tested. This is the essence of the scientific experimental method and presupposes the uniformity of nature. Inductive logic is the product of our intelligence and capacity to observe, describe, and encode natural phenomena in our brain, a view that philosophers and logicians have traditionally dismissed. Drawing valid conclusions is a *natural process* that can be observed in preverbal humans and in animals.

9.1 The early story of logic

The ideas about logic have been changing since its origins, from a practical tool to the development of independent formal systems, which for a while became progressively more abstract. Plato and his followers believed in the existence of *innate ideas*[1] and that some knowledge could be obtained *a priori*. It is said that

---

[1] Humans and animals have many innate abilities (suckling, walking, etc.) that require some practice to make them fully functional. Innate abilities are explainable by preformed circuits that become functional at birth.

Socrates was such a gifted teacher that he could elicit the solution to complex geometrical problems from a young boy without instruction [1]. This, and similar examples of the Socratic method were taken as an indication of *latent* knowledge that could be brought into consciousness by *anamnesis*, that is, by "recalling to memory" ideas acquired during a previous, purer existence. Aristotle disagreed with the idea that logical thinking was a natural process that could be inferred from factual evidence.

Plato's academy did not have a kindergarten and philosophers were not in charge of educating children, so they could not have observed, as Piaget and other psychologists did, how the presumably "innate" Platonic ideas actually originated. Early philosophers did not realize the role of experience in shaping the mind, and they were not aware of implicit learning [2]. The inability to detect implicit learning has led philosophers and mathematicians to look for alternative explanations to account for knowledge which they possessed, while being unable to explain how they had acquired it. Thus, they resorted to auxiliary hypotheses, such as innate ideas, *a priori* knowledge, axioms, etc. The problem with all these ad hoc hypotheses is that they have never been legitimized.

One of the possible explanations for the tendency of logic to progress to abstraction may be the strong influence of Platonic ideas, Kant's ideal of *pure* reason, and the presumed identity between mathematics and logic. That logic could be the formalization of natural reasoning or the symbolization of the structure of nature was completely unacceptable to many logicians. However, the nature of logic can be understood only by studying its natural history and the development of logical thinking in animals and young children. Thus, understanding logic involves studying not only its evolution through history (phylogeny), but also observing its individual development (ontogeny) in children.

We can draw valid inferences without studying logic. Before logic was developed as a science, humans were able to deduce true conclusions from experience and from generally accepted knowledge. Drawing valid conclusions is a *natural biological process* that takes place in the brain. Clearly, *pure logic* had to be abstracted from *natural logic*. If we were not able to identify true or false thoughts and propositions before developing logic, we would have been trapped in an infinite regress, without the capacity to extract logic from natural thoughts.

As an adolescent, I found that reading the history of early logic was deeply rewarding, perhaps because the spirit of inquiry of young minds seems to parallel that of the ancient Greeks, who were obsessed with the search for truth and with the methodology of argumentation and discussion. Since science was still embryonic in ancient Greece, they needed a method to settle controversies and to find truth in philosophy, religion, and law. The dialogs of Plato (427–347 B.C.) illustrate how they used arguments in their search.

The Sophists, in the fifth-century B.C., speculated on theology, metaphysics, and mathematics, as well as on natural and biological sciences. They became well known for their fallacious arguments and paradoxes, and they considered "dialectic"—the art of reasoning or disputation by question and answer—an important component of education. According to Aristotle (384–322 B.C.), dialectic was invented by Zeno of Elea (OED), who founded the Stoic school of philosophy ca. 300 B.C. The analysis of language evolved into the syllogism, a form of *deductive* reasoning that consists in deriving specific conclusions from a major and a minor premise. This indicated that, in its origins, logic was *an empirical tool to find the truth* by using examples and counterexamples as an essential component.

The logical method was further developed by Plato, who used the term *dialectic* to include logic and metaphysics, but its meaning was restricted by Aristotle to the "instrument" of all reasoning and to the method of finding the truth (formal truths) by arguing, as opposed to the demonstrative method of science (material truths). Aristotle considered syllogisms *direct principles* of logic, *negating* the possibility that they could be derived as *inferences* from factual evidence. Thus, for Aristotle, logic was unrelated to natural processes. This was a mistake and perhaps the earliest stance against *psychologism*, which we will now discuss.

Through the Middle Ages and in earlier English use, dialectic was a synonym of logic as applied to formal rhetorical reasoning, logical argumentation, or disputation. Euclid, an earlier Greek geometer, was one of the strongest proponents of the idea that geometry is self-evident (see Chap. 8). In contrast, by the end of the Renaissance, Francis Bacon (1561–1626) instigated the induction-based scientific method that relies on generalization from several specific instances to a more general hypothesis, whose validity can be tested with experiments. Bacon conceived axioms as *empirically derived laws*. David Hume (1711–1776) also believed in the supremacy of the experimental method and in the concept that induction presupposed the uniformity of Nature. All these ideas were important for the formulation of the empirical method by J.S. Mill (1806–1873), who also accepted that mathematical axioms were empirical generalizations [3].

The principles or the "laws of logic" were sometimes identified with the "laws of thought." The idea that there are *natural laws* that rule the universe is metaphysically unacceptable because it complicates ontology by requiring not only the independent existence of laws as such, but also a connection or enforcer that keeps causality working and holds nature to lawful behavior. Laws should be interpreted as empirical *regularities* derived from energy transformations and causal interactions (just "the way in which the cookie crumbles"). Laws cannot control how facts or processes must occur. The empirical origin of laws does not negate their epistemological value. Nobody has to enforce the fact that a crystal glass will break when it falls to a marble floor. Natural laws are formulated by humans to be isomorphic to the transformations of matter-energy under specific circumstances.

This idea originated with Aristotle, and had many illustrious followers. Leibniz (1646–1716) believed that some logical principles were innate, such as the *Principle of Contradiction*, which asserts that no proposition can be both true and false, and the *Principle of Sufficient Reason*, stating that nothing occurs without a reason. But it seems unlikely that anybody could have formulated these two principles without having first experienced the causal interactions we witness every day. Leibniz influenced Kant (1724–1804), who believed that the ideas of time and space were also innate. Leibniz postulated a *characteristica universalis*, an ideal universal symbolism that would allow us to "calculate" truth [4]. The universal language never materialized, but the idea was obviously fascinating, and made an impression on many logicians.

The initial success in symbolizing and formalizing logic probably first suggested the identity of logic and mathematics. This was further elaborated by George Boole (1815–1864), who thought that mental processes *follow* the laws of logic. Thus, by looking at the laws of logic we could find out how the mind works. However, Boole realized that logic could not be conceived as a kind of *physics of thought*, because its laws could be violated whereas the physical laws could not [5]. The identity of the laws of thought and the laws of logic is incorrect, because most of our thinking is not "logical."

The modern scientific and philosophical logic of deduction has become closely allied with mathematics since the work of Gottlob Frege (1906–1978), who made important contributions to the notation and formalization of logic and to the logical foundations of arithmetic [6]. As was extensively covered in Chap. 8, Frege thought that the principles of logic were self-evident, and he tried to *reduce* arithmetic to logic, a conception known as *logicism*. It is important to reiterate that Frege accepted the cooperation between mathematics and philosophy, but strongly objected to any psychological "intromission" [6]. He thought that, ...*psychology should not imagine that it can contribute anything whatsoever to the foundation of arithmetic* (ibid. p. VI). Frege preferred "proofs" derived from his own subjective, self-evident ideas to "any confirmation by induction" (ibid. p. 2). As with mathematics, Frege also vehemently opposed any psychological influence in logic (psychologism). Unfortunately, he never realized that thinking, logically or not, is a physical process.

Frege made a profound impression on several generations of mathematicians and logicians, particularly on Bertrand Russell (1872–1970). Ironically, Russell found a paradox that invalidated Frege's initial formulation, and had to invent the theory of types to patch up Frege's system. Russell and other philosophers tried, but failed, to show how the foundations of mathematics lay in logic. Many logicians have elaborated formal logical systems using symbolic techniques and mathematical methods of deduction. The fact that mathematics follows logical rules does not mean that logic, and logic alone, can constitute the basis of mathematics. Nature is also logical, but this does not imply that we can find the laws of nature in logic. The process is exactly the reverse: logic and mathematics are implicitly learned from Nature. Logic and mathematics are symbolic structures abstracted from the forms and processes of nature; they symbolize its structure and fundamental relationships. Today we know that arithmetic and logic are both *human inventions* that were based on independent intuitions and experiences of natural processes that were further developed using extrapolations and analogies.

9.2 Deductive logic can be formalized

To avoid the ambiguity of words and to facilitate the analysis of logical statements, mathematicians and logicians developed several symbolic and graphical notations.[2] The symbolic notation of logical arguments facilitated operations with complex statements that could otherwise be ambiguous. Formal systems provided a quasi-mathematical language that made it easier to check the accuracy of propositions. Thus, thoughts were first expressed with words that were later replaced by symbols for objects and symbols for operations. In addition, some logicians developed graphical representations that illustrated the physical relationships between the objects and concepts involved (Table 9.1). The symbolic representation of logical statements exhibits a "form" which logicians believe *guarantees* the truth of the statement, providing that the premises are true. Logical forms are evident in our thinking, but they originate in our capacity to encode and model natural forms and processes. However, what logicians, from Aristotle to Frege were not willing to recognize is that logical forms have been *implicitly abstracted* from the structure and processes that take place in Nature. Thus, logic is natural because Nature is logical.

The idea that the formality of logic is independent of the structure of the world and causality is an illusion that has its roots in Aristotle and Descartes. Actually, the causality intrinsic to macroscopic natural processes is what guarantees the validity of our statements. This is shown by the variety of natural processes, which have their own specific relations and create the need for many "logics" with different forms. All systems of formal logic are said to be *a priori*, but this is not the case, because when you are old enough to understand the meaning of *a priori*, you have already implicitly acquired many logical experiences *a posteriori*. Logic, like geometry, uses definitions and rules that were *explicitly selected and modified* to produce the results that agree with our

---

[2] Different symbolic and graphic notations were introduced by mathematicians, such as George Boole, Giuseppe Peano, Leonhard Euler, John Venn, Charles S. Peirce, and others.

Table 9.1 Some elemental logical notations

The use of symbols in formal logic facilitates the analysis of the structure of propositions and the development of propositional and predicate calculi, that are useful in dealing with complex propositions. The *propositional calculus* deals with the mechanics of inferences. In propositional calculus, *operators* ($\sim$, &, $\vee$, $\supset$ and $\equiv$) are used *to substitute for expressions of the natural language*

$\sim$ = negation (not).
& = conjunction (and).
$\vee$ = disjunction (or).
$\supset$ = implication (if…then).
$\equiv$ = equivalent (is equivalent to).

The operators are similar to the "+" and "−" signs of arithmetic, which were also invented to symbolize natural processes, and which are symbolized in logical operations. The use of signs in an equation is a way to represent the operations to which the variables are subjected.

Additional symbols are the *quantifiers*. Variables are limited by prefixing a quantifier such as "all" or "some". "All", the universal quantifier, is written as ($\forall$x) and "some", the existential quantifier as ($\exists$x). There are additional quantifiers to symbolize many, few, more than half, exactly one, etc. There are several equivalent notations as discussed by Haack [33].

The variables, operators and quantifiers are symbols used to formalize expressions, develop several systems of calculus, determine the validity of propositions, formalize the rules of the syllogism, etc. The symbols are also used to express the set of rules for each system, write axioms, or derive theorems, as in mathematics.

The independent systems, such as the propositional calculus, predicate calculus, and natural-number arithmetic are self-consistent, but they are artificially limited, and create barriers rather than unifying knowledge. Efforts to symbolize natural expressions show that, in its origins, symbolic logic was an applied, empirical science, as was mathematics in Mesopotamia and Egypt. There is no doubt that logic was meant to be, not only normative, but also a tool to determine the veracity of propositions and arguments.

intuitions, just as Euclid and Hilbert did with the axioms of geometry, as discussed in Chap. 8.

The neural encoding of signals, symbols, and words is essential for thinking and for communicating. This is supported by our capacity to detect, interpret, and emit signals, which are perhaps the most characteristic functions of the nervous system. Thus, there is no reason to believe that the true or false value of propositions could be established a priori. Scientific experience reinforces the concept that propositions are analyzed by examining statements and arguments that are then judged true or false, on the basis of arguments and existing evidence. This indicates that the formalism of deductive logic, including its premises, have been slowly abstracted from observation of natural processes.

9.3 The problem of induction and the emergence of empiricism

Philosophers and logicians have traditionally concentrated on *deductive* logic, in which the conclusions are *implicit* in the premises. However, induction is the most important side of logic, because it generates new and more comprehensive knowledge. The problem is that induction has resisted all attempts at formal or non-empirical validation.[3] In traditional inductive logic, the process seems to be the opposite of deduction; thus, particular facts are used to infer general premises, but logicians have been unable to create any tight *formal inductive system*. This is because there is no clear way to determine how many observations are necessary to formulate a *general premise* or a universal affirmative proposition. As discussed below, induction and deduction are not reciprocal processes. Classical induction or enumerative induction is known today as *inductive inference*, which is contingent because it can be disproved by a single opposing finding.

The failure of classical logic to produce a credible formal theory of induction convinced logicians during the second half of the twentieth century that logic is actually an empirical science. Despite the opposition that logicians and mathematicians have voiced against empiricism, the roots of all formal truths are derived from experiences. These presumptions are supported by psychological and developmental studies in humans [7] and by the modern approaches to induction developed by computer scientists [8, 9]. Moreover, the need to devise computational inductive systems for expert advice and decision-making has prompted the collaboration of logicians, psychologists, computer scientists, and neuroscientists. This has resulted in remarkable progress and a copious bibliography,[4] but not in a general theory of induction. Despite the lack of formal systems to prove valid *inductions*, the fact remains that major premises serve for valid syllogisms (*deductions*). For example, we can say with confidence (at least during the early twenty-first century) that "all humans are mortal."

As indicated above, Francis Bacon laid the foundations for the scientific method that relied on *induction* to formulate hypotheses. His work was continued by David Hume, who was concerned with causality and particularly with causal or

---

[3] Formal systems exercise a mysterious attraction for logicians and mathematicians, some of whom do not realize that everything is matter-energy. Contrary to Plato's dreams, forms without matter do not exist, because thinking is a physical process.

[4] For recent progress see Refs. [10–15]. In addition, there are several entries in the Stanford Encyclopedia of Philosophy http://plato.stanford.edu/contents.html, and several sites on the WWW.

inductive inference [16]. He believed that experiences foster "habits of the mind" which lead us to anticipate an outcome. Hume's objective was to distinguish good inductive habits from bad. He also believed in the supremacy of the experimental method, and that induction presupposed the uniformity of Nature, which is evident in the universality and regularity of macroscopic processes and in the validity of science. All these ideas were important for the formulation of the empirical method by J.S. Mill, who believed in the law of universal causation according to which inductive inferences were motivated by causal relations and by regularities and similarities. As indicated above, Mill thought that mathematical axioms were *empirical* generalizations.

9.4 The relevance of psychology to logic

Modern psychology developed from studies of the physiology of perception and the psychology of learning, initiated by Herman L. F. Von Helmholtz (1821–1894) and Wilhelm M. Wundt (1832–1920) more than a century ago. These investigations paved the way for experimental approaches to learning. Experimentation is now the default methodology of psychology and indeed of all scientific disciplines. In the first third of the twentieth century, *genetic epistemology* [17] and later *naturalized epistemology* [18] focused on the normal development of logical notions and the acquisition of knowledge. It was unavoidable that experimental psychology would tackle the problems that philosophers and logicians could not solve, namely, the logic of induction.

Logicians have insisted that formal logic is autonomous and should not be confused with the empirical study of reasoning, which is presumably the subject matter of psychology. Any reference to psychology is equally unacceptable to philosophers, logicians, and mathematicians, who seem to fear *psychologism*. This claims that logic is nothing but the psychology of correct thinking, and that mathematics is the natural psychology of mathematical thinking. Psychologism has been a tendency in English philosophy since Hume, even though it has been opposed by many mathematicians, including Frege and Edmund Husserl. As discussed in Chap. 8, Frege was the most obstinate antagonist of the contributions of psychology to logic and arithmetic, and was supported by R. Carnap, who also believed that the gradual elimination of psychologism was one of the most important achievements in the development of modern logic.

Some of Frege's ideas will be reviewed here, because they are equally relevant to both logic and mathematics. Frege believed that logic and psychology look at different properties of the same thing in different ways; he thought that we have to distinguish between the "psychological" laws and the "logical" laws, which are the "laws of valid inference" [19]. The psychological laws of thinking would describe all kinds of thoughts, whereas logical laws would be concerned only with truth. Frege contended that "truth" is independent of psychology and falls in the exclusive domain of logic. It could be said that the *formulation* of a true proposition is a psychological process, but the *verification of the truth* of propositions is the exclusive domain of logic. This is a fallacious conclusion that subdivides and compartmentalizes natural processes to the point that the views of different specialists become unacceptable to each other. Frege believed that what is true is true independently of the people who recognize it. However, what Frege did not realize is that only *a person with a physical brain* (or a man-made algorithm fed into a computer) can formulate or verify the truth of a proposition.

The development of logical thinking as a natural process has been studied by Piaget and collaborators [20]. Piaget was quite aware that logic is implicitly discovered by every child. He compared children to scientists, and believed that every child reinvents for himself the logical rules of science [20] Piaget's genetic (evolutionary) epistemology [17], as well as many of his detailed accounts of the development of logical and mathematical knowledge, cast serious doubts on the wisdom of Frege's exclusion of psychological reality to explain certain types of knowledge.[5] Piaget is not alone; Quine and others have also proposed that epistemology should study the *actual formation* of knowledge. These ideas go back to the early English empiricists and obviously to J.S. Mill.

Another limitation of Frege's ideas is the concentration on deductive systems rather than on *inductive* logic. Classical logicians have constructed gigantic symbolic structures that can prove the truth or the falsity of a proposition, but cannot provide any new knowledge. All deductive systems are derived from a priori principles or "self-evident truths" that must be provided by somebody else, or accepted without a demonstration. However, premises and axioms, and the natural laws, are the product of implicit learning, as discussed earlier for the case of Euclidean geometry. Induction is a complex mental process that has resisted logicians' efforts to formalize it. Traditional logicians could "prove" that Socrates is mortal, but as I will discuss below, they could not prove *logically* the major premise, that *all humans* are mortal, which is an empirical observation. Logicians were concerned mainly with the mechanics of the deduction, which is actually a trivial problem.

In contrast to experimental psychologists, classical logicians have avoided the problem of determining the certainty of the major premises by saying that they are only concerned with the deductive arguments and with the *formal structure of the deductive process*. Logicians and mathematicians accept premises and self-evident axioms *a priori*, as fundamental truths upon which to base their reasoning by thinking, but this is a physical process that we do not fully understand. In an effort to *de-psychologize* logic, Bertrand Russell stated that arguments are valid by virtue of their form, not their content [21]. Russell added that the form of classic syllogisms should be modified to say, "*If* all men are

---

[5] Piaget has been systematically ignored by philosophers. In part, this could be explained by Piaget's use of a psychological language full of unconventional theories and figures of speech.

mortal, and Socrates is a man, then Socrates is mortal". Is it possible that Russell was not sure that all men are mortal? Not really. What he probably objected to was that the major premises (which are inductions) were not logically demonstrable, even if they were considered true statements. Parenthetically, the stoics were the first to state the major premise of their syllogisms by preceding them by the conditional conjunction *if*.

The denial of the relevance of psychological and empirical concepts to philosophical issues is completely unjustified in light of what we now know about the development of logical notions, implicit learning in children, and the physicality of the mind. This denial has contributed to slowing down the progress of philosophy. The essence of philosophical activity is thinking, reasoning, and reflecting, even though some philosophers still forget that all these activities consist in physical brain processes.

During the early twentieth century, deductive logic was transformed from a *practical tool* to an abstract, deductive science relying on symbolic techniques and quasi-mathematical methods. The need to program computers was instrumental in prompting the development of formal languages that try to mimic not only natural processes, but also human activities from mathematics to cognition with an accuracy that our brain does not have.

9.5 The biological roots of induction

The capacity to generalize is *deeply rooted in biology* because fast learners are fitter for survival. Inductive knowledge is swift and mostly unconscious; it does not take too many bites from horseflies to learn that they may hurt you, or that fire burns, and that some foods are unpleasant or make you sick. Thus, the next time we are in the same situation, we will avoid a similar unpleasant experience. Animal species with limited inductive capacity are at a disadvantage for survival, so they must reproduce faster for the species to survive. The emotional value of the stimulus and its survival significance are obviously critical in establishing the strength of the inductive inference and the speed of learning. Despite its high survival value, a generalization made out of one or two experiences cannot be justified, either scientifically or philosophically. However, the mechanisms that make such rapid generalizations are quite efficient and have deep biological roots. Moreover, the same mechanisms of induction operate not only at the basic survival level of popular psychology, but also at all levels of scientific induction. A falling apple, or a novel phenomenon observed once is not enough to formulate a law; but in many cases, it is sufficient to set up a hypothesis that can be tested and retested until it becomes believable.

Inductive learning takes place naturally and efficiently because a considerable fraction of inductive reasoning is acquired like grammar, by implicit learning. This is why some ideas that develop unconsciously are thought to be "innate" or "a priori." Logical thinking has been considered innate by many generations of philosophers, but there is no doubt that the logic used in everyday reasoning is learned by babies and children through interactions with their environment, as described by Piaget and collaborators [7, 22]. Thus, the most elemental forms of inductive learning are common to humans and higher animals. We can also infer that animals use their experience in making decisions, such as avoiding predators or following the pecking order of the group. These elemental forms of induction and deduction cannot be called logic in the classical sense, but there is little doubt that the animal mental processes are similar to the basic processes that take place in our brain. However, we are completely set apart by our superior capacity to analyze ourselves, to abstract and generalize experiences, and to integrate them into coherent knowledge.

The empirical evidence indicates that both *inductive inference and deduction are biological processes* that take place in the brain and have high survival value because they allow us to predict the behavior of prey and predators. The biological roots of inductive inference and deduction should not be ignored when we try to establish the nature of logic. The unconscious character and the smoothness of implicit induction have obscured the origin of many ideas and have fostered the notions of self-evident ideas and a priori knowledge.

Inductive inferences must be studied empirically because (1) they are natural processes and (2) classical logicians have failed to justify them. The negation of the relevance of psychological experiences and of the biological roots of logic is one of the reasons that classical logicians and philosophers have failed to shed light on this problem. To justify the independence of the different specialties, philosophers have fragmented the natural world into domains of different disciplines that are assumed to belong to independent realms. The exclusion of evidence on the grounds that it "belongs to a different discipline" (usually psychology) is one of the primary excuses that some logicians and mathematicians have used to maintain the independence of their disciplines. Unfortunately, the fragmentation of empirical learning precludes the unification of knowledge and hinders cross-fertilization, while also creating apparently irresolvable paradoxes and contradictions and resulting in endless arguments.

9.6 The naturalistic approach to inductive inference

The failure of classical logicians to produce a generally acceptable theory of induction has contributed to today's preference for referring to induction as *inductive inference*. To be meaningful, an inductive inference must (1) be empirically confirmed, and (2) not have opposing instances. These two conditions imply that inductive inferences are *empirically* based, and supported by probabilistic approaches. This also means that they are *contingent*, because they are at risk of error and are sensitive to background information and context. So far we can be sure that "all humans are mortal" because there have been no instances of immortal humans. Thus, inductive inferences rely on assembling rules or models, the complexity of which actually justifies the existence of different specialties and professions.

Inductive inferences are not part of classical logic because they can only be acquired empirically. This amounts to saying that inductive inferences are grounded on experiences and derived from the analogical encoding of the processes and structures of the world. Thus, the creation of knowledge resides in the hands of natural scientists, who formulate the regularities of Nature into "natural laws". As previously indicated, these laws are contingent, because they can be modified to accommodate newly discovered facts. The symbolization of rules and implementation of expert systems in computer algorithms relies mainly on modeling human cognition using a variety of approaches, including neuronal networks and artificial intelligence.

In contrast to the contingency of inductive inferences, the deductive inferences of mathematics are necessarily true because arithmetic and geometry were initially built to be isomorphic to the perceptual structure of the world. The empirical origin of geometry and arithmetic is clear in their elementary forms. The fact that the axioms of Euclidean geometry were derived from experience is quite evident in Hilbert's work [23]. He had to add more than 30 axioms to Euclid's list to justify geometry with what he thought were truly *axiomatic axioms*. Actually, Hilbert unwillingly showed that the Euclidean geometry of flat surfaces was implicitly learned, and that Euclid's five axioms were not sufficient to justify it. Obviously, nobody learns geometry by derivation from axioms. Elementary geometry is implicitly extracted from the geometric environment. Similarly, reason itself—our capacity to think logically—does not develop without experiences. Reason is implicitly developed in dealing with the objects and the processes of the external world. Philosophers have failed to justify the nature of the relationship between reason and world events. They have never explained why reason can deal with Nature or why reason and causality are isomorphic and justify each other. Causality is not a metaphysical invention, but the way in which energy and matter transform into each other and drive natural processes in the macromolecular world.

9.7 The brain encodes models of the world

Implication would thus be a kind of artificial causation in which symbols connected by rules represent events connected by causal interaction . . . .

K.J.W. Craik, *The Nature of Explanation* (1943, p. 63).

One of the fundamental properties of the brain is memory, which makes learning and knowledge possible. Memory, as described in Chap. 7, is a physical process that consists in changes in the strength of the connections between neurons. There are different kinds of memory that involve different parts of the brain; we have the ability to recall objects and complex processes, faces, words and languages, and our own internal thoughts. This information is essential to make implicit inferences about the world, even without explicitly formulating our thoughts; implicit inferences provide animals and humans with rudimentary knowledge to navigate through their environments. Higher animals and humans are able to internalize models of objects and processes from the external world and to predict the behavior of prey and predators and of other natural events. As discussed in Chap. 6, our capacities for language and abstraction give us an enormous advantage over animals; in addition, we can analyze in retrospect patterns of behavior and natural process that allow us to generalize from experiences.

Making inferences from previous experiences, even in the absence of language, is essential for predicting the behavior of living organisms. The development of these mental abilities has preoccupied psychologists for more than a century, and one of their most prolific representatives has been Jean Piaget, whose studies on the intellectual development of children has contributed substantially to understanding, not only knowledge, but also ourselves. Piaget was particularly interested in the development of logic in children, and he studied the way they generalize from experiences, and the way they start to make inferences. Children classify objects and organize them in series according to similarities and differences. For example, they learn that wooden objects float and that other objects sink, and that metal objects do not break when they fall, whereas glass breaks. Thus, children not only learn about the properties of objects, but also about their relations and interactions. They also discover causality, which they easily link to their own actions, such as pulling, pushing, moving, and dropping. Besides, they learn that the actions of other people produce similar results. Children's capacity to classify and to seriate objects, and to infer causality, has been equated with the discovery of an elemental logic [7].

Modern studies on development exploit more subtle detection methods than those used by Piaget and his collaborators, and they infer their results, not from the explicit responses of children, but from their implicit tendency to look less and less at familiar events and longer at novel events. Thus, recent studies indicate that causal relationships can be perceived by infants at 6.5 months of age, much earlier than Piaget suggested [24]. However, to be aware of causality requires the perception of several "frames" that depict the process, before and after the causal agent intervenes. For example, to conclude that metal objects do not break when they fall is a logical inference that requires more than one observation. Piaget compared children to early scientists, in that their development roughly recapitulates the development of science. Thus, as

Craik [25] indicated, events connected by causal interactions become predictable.

The naturalistic approach of cognitive scientists is at odds with the Platonic hopes of early logicians, who assumed that logic could be derived *a priori* from some kind of abstract reality. There seems to be no doubt that logic is a natural inference derived from the observation of causal interactions. If we imagine that a malicious spirit could suppress causality, nothing would remain the same. The duration of days and nights would be irregular, clocks would become meaningless, and objects would change into other objects and move in all directions or not at all. Everything would be chaotic and unpredictable, and logic would have no meaning. The determinism and the regularity of Nature are symbolically encoded in our thinking as a formalism, which we interpret as a *logical determinism or an implication*. In other words, we unconsciously encode the determinism of the external world as logical implications. Under a malicious spell, the premises and the conclusions of syllogisms and truths of reason would be meaningless. The predictability and the *certainty* that thinking logically provides *emerges from the determinism of Nature, and not from logic itself.* Thus, natural processes *become* the underpinnings of logical thinking.

We can generalize and say, *it is natural, therefore it is logical.* Thus, *necessity or implication is to logic what causality is to the macroscopic physical world.* However, the idea of causality is not innate but learned [7]. The first law of thermodynamics is the law of conservation of energy. It implicitly formulates the law of causality by stating that energy does not vanish, or arise out of nothing. This is why the position and the mass of objects—or the shape and meaning of our logical symbols—will not change, unless they are affected by a source of energy; the first law also underpins Leibniz's *Principle of Sufficient Reason* (which states that nothing occurs without a cause). The law of conservation of energy was put forward in a progressive fashion by James Prescott Joule (1818–1889), Julius Robert von Mayer (1814–1878), and Hermann von Helmholtz (1821–1894). The first law of thermodynamics should be incorporated into any respectable system of logical axioms. Otherwise, the Aristotelian "laws of thought" (identity, excluded middle, and non-contradiction) would vanish together with the logical principles and all knowledge. The notions of identity or difference exist only if the law of conservation of mass-energy is valid. All this implies that logical thinking and reasoning are direct consequences of the brain encoding the structures and processes of our world. We know today that our thoughts are *physically encoded* in our brains, so they do not belong to a Platonic realm; instead, they are parts of our own physical self, providing a way to encode knowledge and to understand the world. Thus, Aristotle's formal truths and the truths of our thoughts derive from the material truths that we observe and implicitly internalize.

9.8 Reasoning and logical thinking are physical processes

Reasoning is a biological process by which our percepts are organized to understand the environment. The sensations produced by objects and processes from the external world are internalized as encoded neural messages and are analyzed in different parts of the brain as sounds, images, and complex events. The internalized models allow us to make predictions and to have expectations, which constitute an implicit form of logical thinking. Most internalized objects and processes are presumably encoded in analogue brain circuits. However, the degree of isomorphism of the encoding is variable and is dependent on the qualities of the objects internalized. For example, we can safely assume that triangles or the space that surrounds us is encoded almost isomorphically, whereas what we perceive as colors, sounds, and smells is not. Qualitative experiences are neural surrogates that *represent*, but cannot portray the situation, as it is (Chap. 5 and Musacchio [26]). Even so, our percepts are sufficiently representative for us to survive and prosper.

The physicality of thinking has been demonstrated many times in a variety of ways. For example, we have known for centuries that brain diseases, opiates, and alcohol interfere with thinking. Mapping studies of brain lesions have provided the earliest reliable demonstrations that cognition depends on the function of certain brain regions. It is particularly important that lesions of the frontal lobes interfere with our abstract thinking and with our capacity to make inferences and analogies, which are basic forms of logical thinking. The most convincing demonstrations of the physical character of brain functions are those in which the experimental subjects are given specific tasks while they are subjected to brain imaging studies. For example, the anterior part of the frontal lobes (prefrontal cortex) has been repeatedly shown to be involved in inductive and deductive reasoning, but deduction tends to activate the left inferior frontal gyrus, whereas induction activates the left dorsolateral prefrontal gyrus [27–31] (Chap. 4 and Fig. 4.2).

Additional tasks recruit different areas of the brain; the left temporal area is recruited in semantic or linguistic reasoning, and the memory-associated hippocampal system is recruited in reasoning about the spatial environment. Thus, reasoning is dependent on activation of brain regions most involved in abstract thinking, but additional brain areas are recruited to deal with semantic, spatial, or arithmetical problems. Even though we still do not know in detail the physical language used by the brain to encode different aspects of reality, there is no doubt this involves activation of neuronal networks within these brain regions. For example, memory and learning consist of changes in the connectivity and activity of neuronal networks within specific brain regions with a special microscopic organization to encode the different kinds of information received from sensory inputs, as described in Chap. 7.

Thinking involves processing the encoded models and recombining them to produce new thoughts, which is a rearrangement of previously internalized processes. As we have learned from our children, thinking logically is an acquired ability, like walking or cycling; it requires training and reinforcement to establish and maintain the proper connections that will associate symbols in a meaningful fashion. The brain learns to reason, so in analogy to the learning and development of other brain functions, logical thinking implies the development of self-correcting circuits that encode the logical operations postulated by logicians, such as those of formal logic. This process—like the learning of all skills—requires experiences for development and tune up. Most likely, our logical circuits are constantly formed, tested, and modified throughout development. The mind-brain complex develops slowly from before birth to adolescence and the early twenties, but some of its basic functions, such as certain reflexes and simple behavioral patterns are genetically determined and hardwired at birth. However, the nerve fibers establish contacts with their final targets by trial and error correction. The target cells require functional stimulation and provide growth factors and cell-surface signaling molecules that are essential for the survival of individual cells and their connections. The establishment and consolidation of the connections require—besides functional stimulation—appropriate levels of nutrients and hormones.

The physical world is encoded in our brains through sensations, experiences, and intuitions, so when the brain encodes a valid analogy between our expectations and the actual world, we find what we call *truth*. Valid statements are *symbolically isomorphic* to some aspect of reality. Language is only a contingent link between the objects and processes in the external world and the internalized models that we have constructed in our brains. However, language not only increases our capacity to communicate, but also provides algorithms to think and model the subtleties of the external world. Language becomes a shorthand for thinking more effectively. A language-independent logic of encoded intuitions is used for thinking before we can find the appropriate expressions; language-independent thinking is also used by preverbal humans and higher animals to control their own behavior.

Logical and mathematical operations are models of physical processes. Additions and subtractions did not originate in "pure" mathematical thoughts, but obviously in everyday life, during which our ancestors gained or lost something. Multiplication and division are more abstract, and perhaps should be viewed as higher order symbolic operations, in which addition and subtraction are combined.

Reasoning, thinking, and explaining are complex brain functions that are important to understand and solve problems through analogies that mimic natural events. In keeping with many of the pioneering ideas of Craik, neural circuits must *model* the way in which the external objects interact, following the causal laws of Nature [25, 32]. In other words, the brain circuits that handle relations between objects and processes encode the processes of the external world. The brain does not produce internal images that are visible as if they were pictures, but the encoding is interpreted *as if it were reality itself*. The activity of brain circuits is what implements everything we perceive. Thus, all perceptions are illusory in the sense that the elements of the external world never are directly internalized as such (Chap. 5). We only perceive the analogical models and qualitative experiences that filter through the senses as encoded neural signals. However, since we are only neural processes, the neural signals get perfectly integrated into ourselves.

Logical operations are fundamental processes of the cognitive machinery that are modeled and programmed during learning and development. However, most organisms are born with some prewired neurons which are tuned up after birth by experience, like sucking, eating, or the ability to walk, which fully develop only through trial and error correction. Innate cognitive abilities are rarely completely functional in humans at birth, but there are no doubts that the capacity to develop certain brain circuits is genetically determined.

9.9 The emergence of different logics

Logic discriminates valid from invalid arguments, but the details of arguments vary with some subjects and between disciplines. The complexity and intricacy of Nature and culture indicate that a simple logic does not fit all aspects of reality. Thus, many practitioners have created logics out of their own disciplines. Moreover, the evolution of culture introduces new perspectives which, because of their increasing complexity, require additional extensions in the way we argue and construct inductive and deductive inferences, whence they can be viewed as different logics. Even though the development of formal logical systems increases the accuracy of logical statements, many specialties thrive on informal arguments during the development of their disciplines. Haack [33] has catalogued 5 main types and about 15 subtypes of "logics".

The logic used by preverbal children and animals is characterized by being language-independent. Studies in infants have shown that they can perceive when something unusual or illogical happens in their environment, as when objects do not obey gravity. Animal logic has also been inferred by observing the cognitive abilities of animals; some animals can solve problems related to navigation, social status, and number comprehension in the absence of a verbal language. Social intelligence is essential to many animals that are able to infer their standing in the pecking order of the group; for example, pinyon jays infer their social status in a group from previous experiences. Tool using and tool making have also been recognized as a problem-solving ability to reach food in animals as diverse as crows, chimpanzees, and bonobos. Some of these animals can also understand numerosity by subitation with four and up to six objects.

9.10 Concluding remarks

Logic—explicit logic—was formalized by the ancient Greeks, even though its implicit forms are individually discovered by every child. Indeed, any child will show surprise when his or her expectations are not fulfilled. Thus, logic is the product of our intelligence and capacity to observe, describe, and encode natural phenomena. Like science, arithmetic, and geometry, logic as a discipline does not exist independently of human intelligence and language, even if it is built into nature. The purpose of logic is to discriminate between valid and invalid arguments [33]. The variety of logics described by philosophers originated in the valid arguments characteristic of different disciplines, indicating that the different conceptions of logic cannot be described under a narrow umbrella.

The logical laws—"the laws of truth"—do not have an independent existence; they have been abstracted by humans from valid descriptions of natural processes. Logic is a human discovery that was born by observing nature. Initially, logic was misjudged as an abstract entity, independent from the mental and the concrete. Frege was the main expositor of the interpretation of logic as if it were the "ethics of thought." He assumed the existence of three independent realms, the abstract, the mental, and the concrete. This view does not explain how it is possible that abstract logical laws could rule natural processes and the truthfulness of human thinking about concrete objects. The abstract, conceived as a non-physical realm, does not exist and cannot have causal efficacy. Actually, there is no evidence that anything nonphysical exists.

We know that concrete objects always "obey" thermodynamics and causality, but nobody has provided the reasons (*sufficient causes*) for logic to rule the putative mental and concrete realms. To explain these relationships, a higher set of bridge laws would be required to connect the abstract logico-mathematical realm with the concrete and mental realms. The existence of hypothetical bridge laws cannot be demonstrated. Moreover, it has a strong Platonic flavor and other metaphysical implications that make this interpretation farfetched and unacceptable. The simplest interpretation is that we abstracted logic from the processes and structures of the world, which is ruled by energy transformations that implement macroscopic causality. We thus abstracted logic from transformations of matter-energy and encoded them in the physical processes of our brain. All these considerations indicate that the formalism that we find in logic has actually been implicitly abstracted from natural objects and processes.

References

1. Russell J. The Acquisition of Knowledge. 1st. ed. New York: St. Martin's Press; 1978.
2. Reber AS, Walkenfeld FF, Hernstadt R. Implicit and explicit learning—Individual differences and IQ. Journal of Experimental Psychology: Learning Memory and Cognition 1991;17(5):888–896.
3. Mill JS. A System of Logic. Eighth reprinted ed. London: Longmans, Green and Co.; 1911.
4. Leibniz GW. Philosophical Writings. London: Dent; 1973.
5. Glymour C. Thinking Things Through. An introduction to philosophical issues and achievements. 1st ed. Cambridge, MA: The MIT Press; 1992.
6. Frege G. The Foundations of Arithmetic. A logico-mathematical enquiry into the concept of number. 2nd ed. Oxford: Basil Blackwell; 1974.
7. Inhelder B, Piaget J. The Early Growth of Logic in the Child. Classification and Seriation. New York: Humanities Press; 1970.
8. Harnad S. Category induction and representation. In: Harnad S, editor. Categorical Perception: The groundwork of cognition. x, 599 pp.: 1987:535–565.
9. Harnad S. Categorical Perception: The groundwork of cognition. (1987). x, 599 pp.: 1987.
10. Polk TA, Seifert CM. Cognitive Modeling. Cambridge, MA: The MIT Press; 2002.
11. Holland JH, Holyoak KJ, Nisbet RE, Thagard PR. Induction. Processes of Inference, Learning and Discovery. Cambridge, MA: The MIT Press; 1986.
12. Churchland PM. The Engine of Reason, the Seat of the Soul: A philosophical journey into the brain. 1st. ed. Cambridge, MA, US: MIT Press; 1995.
13. Churchland PM. A Neurocomputational Perspective: The nature of mind and the structure of science. 1st ed. Cambridge, MA: The MIT Press; 1989.
14. Churchland PS, Sejnowski TJ. The Computational Brain. Cambridge, MA, MIT Press; 1992.
15. O'Reilly RC, Munakata Y. Computational Explanations in Cognitive Neuroscience. Understanding the mind by simulating the brain. First ed. Cambridge, MA: The MIT Press; 2000.
16. Vickers J. The Problem of Induction. Stanford Encyclopedia of Philosophy, 2006 (http://plato.stanford.edu/entries/induction-problem/).
17. Piaget J. Genetic Epistemology. First ed. New York: Columbia University Press; 1970.
18. Quine WV. Epistemology naturalized. In: Sosa E, Kim J, editors. Epistemology: An Anthology. First ed. Malden, MA: Blackwell Publishers, Inc.; 2000:292–313.
19. Carl W. Frege's Theory of Sense and Reference. It's origin and scope. First ed. Cambridge, UK: Cambridge University Press; 1994.
20. Cohen D. Piaget: Critique and Reassessment. 1st ed. New York: St. Martin's Press; 1983.
21. Russell B. Introduction to Mathematical Philosophy. First ed, 10th impression. ed. London: George Allen and Unwin Ltd.; 1919.
22. Kitchener RF. Piaget's Theory of Knowledge. Genetic epistemology & scientific reason. First ed. New Haven: Yale University Press; 1986.
23. Hilbert D. Foundations of Geometry. First ed. Open Court; 1997.
24. Kellman PJ, Arteberry ME. The Cradle of Knowledge. First ed. Cambridge, MA: The MIT Press; 1998.
25. Craik KJW. The Nature of Explanation. 2nd ed. Cambridge: Cambridge University Press; 1943.
26. Musacchio JM. The ineffability of qualia and the word-anchoring problem. Language Sciences 2005;27(4):403–435.
27. Koechlin E, Basso G, Pietrini P, Panzer S, Grafman J. The role of the anterior prefrontal cortex in human cognition. Nature 1999;399(6732):148–151.
28. Koechlin E, Ody C, Kouneiher F. The architecture of cognitive control in the human prefrontal cortex. Science 2003;302(5648):1181–1185.
29. Fuster JM. The Prefrontal Cortex. Fourth ed. London, UK: Academic Press; 2011.

30. Fuster JM. The cortical substrate of general intelligence. Cortex 2005;41(2):228–229.
31. Goel V, Dolan RJ. Differential involvement of left prefrontal cortex in inductive and deductive reasoning. Cognition 2004;93(3):B109–B121.
32. Craik KJW. The Nature of Psychology. 1st. ed. Cambridge: The Syndics of the Cambridge University Press; 1966.
33. Haack S. Philosophy of Logics. Cambridge: Cambridge University Press; 1978.

10 Faith and the validation of beliefs

Summary Faith, in one sense, can generate testable hypotheses which, once empirically verified, are essential for scientific progress. In contrast, religious faith cannot be tested, so it has no truth-value and its conclusions are incompatible with science. Theologians have been unable to formulate the criteria to find truth in religion, because religions are unverifiable. Religions are also contradictory, because they have been invented by different cultures. Religious fundamentalists and right-wing politicians support the teaching of the independence of religion from science. Living in a group requires rules to keep harmony, so there is little doubt that our moral sense originated not from religious faith, but from adapting to live in larger groups. Religious belief is generally not a free choice, but a consequence of early childhood indoctrination. Stressing religious faith as the way to find spiritual and supernatural truths is the most effective way to introduce schizodoxic beliefs in society. Moreover, extreme religious faith promotes the development of sects and cults that produce serious personal and social confrontations.

10.1 Faith has many meanings

Faith is the almost universal way to validate beliefs, but it is also the most ineffective and misleading practice and is unable to discriminate between true or false beliefs. As a way to reaffirm religious beliefs, faith contrasts sharply with the empirical approaches of science and the rational approaches of philosophy. Faith is an inner conviction that equally well characterizes the Catholic priest, the Eskimo shaman, and the African witch doctor. However, faith has several meanings that must be carefully analyzed to avoid misunderstandings. In *common language*, faith is the confident belief in the truth, value, or trustworthiness of a person, an idea, or an object. Faith can also mean a belief in or a loyalty toward a person or cause. However, in the *theological sense*, religious faith is a virtue defined as a secure belief in God, which is understood as a

J.M. Musacchio, *Contradictions: Neuroscience and Religion*,
Springer Praxis Books, Popular Science, DOI 10.1007/978-3-642-27198-4_10
© Springer-Verlag Berlin Heidelberg 2012

trusting acceptance of God's will, without necessity of logical proof or empirical evidence.

The confusion between faith in religion and faith in science or in medicine is misleading, because faith in different subjects is based on evidence of a different nature. *Faith in religion* is not supported by any empirical data, in stark contrast to faith in science and technology. For example, I can say that when I cross the Brooklyn Bridge, I have faith that the bridge will not collapse under my weight. I believe that this is true because the bridge was designed to support many times my weight, and has been tested millions of times since it was built. Thus, instead of blind faith, there is concrete, empirical evidence that the bridge is not likely to collapse under my weight. Similarly, I have *faith in my friend* Otto. I am sure that he will be there when I need him because he has never failed to do so over the last 50 years, since we were both young students. Thus, this faith is backed by a long history of multiple experiences, not by blind faith. I also have faith that in the near future we will be able to cure many more cancers than today, because there is a historical trend and several breakthroughs have been made in that direction.

Philosophers and epistemologists usually avoid discussing the meaning and value of faith, because they consider this to belong to the *philosophy of religion*. This keeps philosophers out of trouble and preserves the idea that faith is an intrinsically adequate method of choosing and accepting religion. Philosophy is customarily divided into different branches to avoid overlaps and omissions with teaching assignments, but a didactic division should not be interpreted as corresponding to any real division in different realms.

There are many expectations in which we have faith, but in which faith alone may not be able to predict events in the absence of empirical knowledge. I can have faith that the stock market will keep going up, or that baseball team A will defeat team B, etc. These are educated guesses that require empirical knowledge, such as knowledge of market indicators or the statistics of previous games in which team A has usually defeated team B. Thus, sometimes when we use the word faith, we actually imply that we have *empirical knowledge or rational belief*, rather than *blind faith*. Keeping all the different meanings of faith strictly separate is important because not all faiths have identical truth-value. Thus, faith derived from previous empirical experiences or scientific evidence has greater probability of being right than faith that is not supported by any such evidence.

10.2 There are inherent contradictions in religious faith

Religious faith is a way to justify beliefs in matters that otherwise cannot be supported, in marked contrast to faith based on empirical knowledge. Adler [1] indicates that, according to Augustine, the essence of religious faith is to be *beyond* proof. The term *beyond* seems to imply that faith is an *infallible* way of knowing. The problem is that all different religions claim to be true, but end up contradicting each other. Religious faith seems to be a naïve and dogmatic way of gaining assurance, which appeared early in the history of civilization, before reason and empirical knowledge became prevalent. In essence, faith is a primeval way of justifying personal beliefs. Dogmatic believers can convince themselves and indoctrinate others, including their innocent children, about something that they cannot demonstrate in any way (see Chap. 3). These believers feel that they can prove their point just by quoting their sacred books—the Bible in this case: "The righteous live by their faith" (Isa. 2.4). "Everything is possible to one who believes" (Mark 9, 23; 16, 17). "Peter, next to Jesus was the greatest Christian of all time because his belief was unqualified and unlimited" (St. Peter). There seems to be no awareness that we could have faith in the "wrong" religion, because early indoctrination is implemented before children have mature judgment.

Religious faith generates contradictory opinions because faith cannot demonstrate the truth of any particular religion or the existence of any specific god. Many predictions about the arrival of the Messiah or the Second Coming of Christ have never been fulfilled. There is ample evidence that the use of blind faith, without prior empirical experience or reasons, to make concrete predictions just does not work. Preachers and theologians try to associate *faith*, in the religious sense, to situations in which faith is supported by empirical knowledge. For example, having faith that next August will be hot in New York City is identified with having faith in the Second Coming of Jesus Christ. This is a non sequitur, and reminds me of a teacher who used to say that, if we had faith that our mother was actually our mother, we should also have faith in God. The rhetorical substitution of one *sense* of a word, faith in this case, by another is unacceptable in the pursuit of truth.

The contradictory beliefs produced by faith have been a problem for theologians, who have never been able to prove the validity of any single religion. For example, the theologian and philosopher Paul J. Tillich (1886–1965) was aware that "[Faith] confuses, misleads, creates alternately skepticism and fanaticism, intellectual resistance and emotional surrender, rejection of genuine religion and subjection to substitutes" [2]. Tillich writes that he tries in his essay to reinterpret the word faith, to remove the confusing and distorting connotations that have been attributed to faith through the ages. According to him, faith has the character of the *ultimate concern* and is an act of the *personality as a whole*. Faith gives certainty, but *it has an element of uncertainty* that requires *courage* to accept it, because there is a *risk* involved. This idea is similar to that of the Danish religious philosopher Kierkegaard (1813–1855), who believed that, without risk, there is no faith, and that the greater the risk, the greater the faith.

This way of thinking and pursuing the truth can hardly be considered an example of acceptable logic. Tillich could not come up with a simple, understandable definition of religious faith, and he needed 127 pages to explain what faith is and is not. However, some of his text includes contradictory statements, both of which are supposedly true, thus making the point impossible to grasp. For example, "[Faith] is infinitely variable and always remains the same", etc. In another nonsensical phrase, he writes, "Faith stands upon itself and justifies itself against those who attack it, because they can attack it only in the name of another faith" [2]. His language is mystical, allegorical, and cryptic, and he does not succeed in explaining how the use of faith to validate beliefs can find truth in religion. Tillich's verbose rhetoric and lack of logic are appalling, to say the least, but this may be the very secret of his success: no scholar has had the time to take him seriously enough to write another book in rebuttal. He sounds like a preacher, but he cannot explain the essence of religious faith. The fact that he has been taken seriously by many is incomprehensible, and certainly does not say much for the intellectual caliber of such theologians.

Faith has produced hundreds of religions, most of which are extinct today. The essence of religious faith and the seeds of its demise are its subjectivity and overriding inner convictions, which inevitably result in contradictory beliefs. Neither faith nor reason can establish the *truth* of religious statements, or determine which of the countless religious faiths is right or wrong. Thus, the problems start when other ideas and competing doctrines are introduced into a closed community, and some members of religious institutions use the new faith to question some Official Faith. The rebels become heretics, and history shows that heretics have been crucified, thrown in the fire, or tortured until they recover the "True Faith" or die.

The overriding importance of faith to the early Christians indicates that the "miracles" attributed to Jesus were most likely "witnessed" by only a few. If 10,000 people had seen the Virgin Mary ascending to Heaven, there would not have been any problem in believing in the Ascension. Clearly, even in biblical times people were skeptical and resistant to belief in second-hand stories. Thus, stressing the value of faith was given primary importance by the prophets, because it was essential for the survival of the new religion. The words of the prophets and the need to explain the origin of everything easily convinced peasants that God created heaven, the earth, animals, and plants, and then made man in his image. These stories were probably reassuring, but today they sound very naive. The complexity of the Universe and the biological world was unimaginable a few thousand years ago. Today, we are more skeptical and realize that faith without empirical or logical evidence cannot determine the truth or falsity of any statement.

It is hard to understand how religious faith is expected to provide true knowledge. Is the existence of a divine being perceptible through faith, even if faith does not work in any other situation? The plurality of religions demonstrates that faith, as a way to know and to search for truth, cannot identify which one—if any—is the True Religion.

Faith cannot be used to prove whether God exists or not. However, let us pretend —for the sake of the argument—that God is necessary to explain the existence of matter and energy. The next problem is to determine whether God is anthropomorphic, that is, a personal god that cares about humans (theism), or only a "force" (deism) or a collection of laws that control the transformation of matter and energy. It is hard to see how these problems could be solved by faith, but again, let us assume that there is a Personal God. The next question is equally hard to answer: Which of the several postulated gods is the Real God? Clearly, the existence of a multitude of incompatible religions seems to indicate that faith, as a cognitive method, cannot *identify either the True God or the True Religion*. Moreover, the situation keeps getting worse, because the varieties of religions keep multiplying, and churches and sects continue to split off from the larger denominations.

10.3 Childhood indoctrination

Our beliefs are quite probably a direct reflection of early indoctrination. Individuals are most likely indoctrinated during the early stages of their moral and intellectual development. Although dogmatic, such indoctrination would appear acceptable, because revered by one's peers. An individual could thus be programmed to become anything imaginable, from a Tibetan monk to a kamikaze pilot. The power of indoctrination is illustrated by the willingness of individuals to die for contingent entities, such as religions and countries. The depth of the convictions produced by early indoctrination explains why some individuals cannot get rid of their early religious beliefs, even after learning the objective truths and the empirical methods of science.

Despite the countless contradictions produced by faith, religions permeate our culture. Most of us have had some religious education at home, from bedside stories to explicit answers to our questions. We may also have been sent to churches or temples for specific religious indoctrination, or to religious schools that teach religion as part of their curriculum. Thus, we were essentially born *into* a particular religion or cult, without realizing until many years later that we were not given a choice (see Chap. 3). Our religious beliefs were therefore determined by the *faith of our ancestors*, many of whom could not be seen as particularly knowledgeable or wise, if we look far enough back. Most individuals, however, will never question their parents' choice, because they have been irreversibly imprinted before they become aware of the imprinting, and only occasionally switch to different religions, usually of similar denominations. Besides, almost everybody will agree that parents have an inalienable right—no matter how ignorant they or their ancestors were—to decide how to educate their children; thus, early religious indoctrination is never questioned.

The reasons for being religious are taken for granted, without any need for justification. Having been born into a Catholic family, I was told many times during childhood that I should have faith in the teachings of the Roman Catholic Church: trust faith, because "it is by faith that you know everything, including who your parents are, etc." Other times I was exhorted to believe "just in case", because "there is nothing to lose but a lot to gain if God exists," etc. Many years later, I learned that Blaise Pascal (1623–1662), the French philosopher and mathematician also advocated a similar hypocritical attitude.

Thanks to the mobility of my family, I had the good fortune not to be regularly exposed to religion until I was almost 8 or 9 years old. As a consequence, my own indoctrination was rather superficial, and religion under such conditions is no match against plain common sense. Indeed, the plurality of religions made me wonder about how I could be sure about any of them. Much later, I realized that the problem of certainty in religion has never been solved: we cannot test whether a statement about a doctrine is true or false, partly because religious beliefs are not verifiable and partly because they are largely based on unreliable books that give contradictory statements in different chapters. Moreover, the variety of religions and religious experiences is staggering, since many of the early prophets described entirely different events. If the same phenomenon occurred in science or medicine, this would be like three different physicians diagnosing three different, but potentially fatal diseases in your sick child, and recommending three completely different treatments. Which one would you choose? How would you decide how to treat your child? Two of the doctors must be wrong, and maybe all three of them. Fortunately, this situation is rarely encountered because medicine is rapidly becoming a science, and an accurate diagnosis is generally the rule. In most cases, bacterial cultures, blood tests, X rays, and image analysis provide a reliable answer. The choice between three incompatible diagnoses is comparable to the choice of a true religion, except that the latter is a choice you have to make in the absence of any diagnostic test.

Thus, there is consensus that religion—if it is going to be second nature—*has to be taught while the children are very young,* beginning with toddlers. This is the reason why all major religions support day care centers, kindergartens, and elementary schools. The religious right-wing supports school prayers in state-run schools and a voucher system so that the government shares the costs for sending children to a religious school. Religious explanations can be readily taught to small children, because religious stories were conceived by primitive and naive societies, long before philosophy and science appeared. The simplicity of religious stories makes religious teaching more comprehensible to the plastic minds of developing children. By contrast, rationally taught science is often beyond their comprehension. Science requires mental abilities that develop much later; the frontal lobes are not fully developed until the early twenties [3]. Teaching science also requires qualified teachers, whereas religious stories can be told at bedtime by well wishing parents who heard the same stories when they themselves were young children.

To promote religion, religious groups—particularly the religious right-wing—also object to the teaching of self-esteem and self-reliance at elementary school. The rationale is that self-reliance could become an obstacle, because children should rely on their parents and teachers, not on themselves. There are certain abilities that are well known to be learned faster and better, or only at a very early age. This is the case of learning to speak a foreign language without an accent, hitting a ball with a bat, or playing the violin reasonably well. Learning at an early age deeply modifies the brain. We have all personally experienced and observed in others that early learning lasts practically forever. Songs, fables, myths, and religious beliefs can be imprinted in such a way that they are not forgotten, unless memories are physically destroyed by a brain disease. Early religious education is so powerful that most individuals never have the possibility of choosing whether to be religious, but assert that *they have chosen to be religious*, indicating that early indoctrination is imperceptible and practically irreversible.

10.4 Mystical and trance-like experiences are delusional

Mysticism is usually the attempt to achieve a "personal union" with God, or some other divine being or with a universal principle. Mysticism implies the belief in the existence of realities beyond perceptual or intellectual apprehension that are central to being, and directly accessible by subjective experience. Mystics apparently fail to recognize that neither the *inner conviction* of an experience nor the *feeling of absolute certainty* are guarantees of the veracity of the experience. I became aware of how misleading certainty could be during my scientific career, when I realized several times that some of my theories were wrong, even though initially I was convinced of their validity. Some theories initially make a lot of sense, explain several facts, and even have a unique intrinsic beauty. However, on further testing, when the predictions of the theory do not hold, the certainty can be rapidly replaced by a deep disappointment. Thus, a *strong inner conviction* of an untested theory is not a criterion of being correct, just as certitude with regard to a hallucination is not a guarantee of its truthfulness. The truthfulness of scientific theories can be tested, whereas mystical and religious convictions are *unverifiable*.

Many trance-like experiences are described in mysterious and inscrutable words that make one wonder about the sanity of the mediums who announce them. Esoteric language with occult meanings has a deep appeal for those in search of revelations, whereas it is just an incomprehensible collection of mysterious-sounding words devoid of unambiguous meaning. For example, mystics describe many weird experiences, such as: the feeling of transcending human consciousness, experiencing the presence of God, sensing being at one with the divine, a feeling of absolute dependence, experiencing the transcendent and mysterious reality of the universe, communicating with a world beyond normal experience, etc. The nature of these experiences seems akin to the experiences and feelings reported by some schizophrenic patients. Mysticism is a form of avoidance, a way to escape the harsh realities of a hostile environment that cannot be controlled [4]. Mysticism in this age seems like a form of adult autism, an abnormal introversion with egocentricity, and an escape into a fantasy that for some is more acceptable than reality.

Supernatural beliefs are not exclusive to primitive cultures, but are instead manifested during the early stages of development of many *individuals in almost any culture*. Supernatural explanations are an attempt to understand and explain the confusing reality before we can comprehend it through reason and empirical knowledge. Supernatural explanations seem to be a temporary solution allowing us to get on with life before true knowledge can be obtained. This means that ideally *supernatural beliefs persist only in the absence of scientific explanations*. It implies that the "primitive mentality," attributed to some cultures, is actually *a developmental stage of each individual*. In primitive cultures, the production of elaborate and farfetched supernatural explanations increases in the absence of satisfactory scientific explanations. These explanations are perfected by successive generations, who enrich the traditions and mythology of their culture. Thus, supernatural beliefs in primitive cultures are maintained or embellished, whereas in higher cultures, these beliefs are arrested by scientific knowledge, while culture continues to evolve in other directions.

The staggering number of religions precludes a detailed examination of all supernatural beliefs. However, even if we could examine them all, we would be unable to determine the superiority or truthfulness of any particular religion, because *religious truth is not verifiable*. At any rate, it is reasonable to believe that a more evolved religion—one appearing in a higher culture—is "superior" to one that requires human or animal sacrifices. The fact that religious beliefs evolve and become obsolete indicates that these beliefs are human inventions, made to suit temporary social or political needs. An omniscient God would have given us a timeless, eternal religion that would not need to be reinterpreted on the basis of evolving cultural and scientific progress. Thus, we can surmise that religions were created not by God, but by human stories, designed to keep us submissive and waiting for Godot, for an eternity.

10.5 Science and mathematics can be validated

Validation of nonreligious beliefs is essential for achieving certainty, but different beliefs require different forms of validation. The validation of scientific theories usually requires additional procedures, such as empirical tests, mathematical and/or computer modeling, supplemental hypotheses, prediction and verification of new phenomena, etc. For example, the theory of relativity was given a vote of confidence when one of its predictions was observed, namely, the bending of light rays in the presence of gravitation. Of course, this left the door open for the construction of even more general theories. Thus, empirical knowledge is a complex, self-correcting, and transcultural form of knowledge that has been developed fairly recently in the history of humanity. However, in the initial stages of a scientific development, we need intuitions or a preliminary hypothesis to design the proper experiments or to find the appropriate proofs. Similar phenomena are observed every day in medicine and in most scientific disciplines.

Scientific knowledge is still incomplete, but it is cumulative and self-correcting. For example, the detailed internal structure of atoms and genes were not known when they were first postulated to exist. This means that scientific knowledge is perfectible. The same can be said about the gaps in the fossil record of evolution, or in any other scientific subject. Obviously, we are certain about the evolution of species, even if some links and some of the mechanisms involved are still missing, and some others might never be discovered. Scientific truth is not relative, but sometimes our statements are ambiguous. For example, the effectiveness of a certain antibiotic for the treatment of diarrhea in the East does not necessarily guarantee similar effects in the West, because the antibiotic sensitivity of the bacteria that cause an apparently identical disease could be different in the two regions. Thus, it is essential to be specific and to keep our statements within the boundaries imposed by the facts that have been verified.

Mathematics, which epitomizes propositional knowledge, must also be validated. It requires intuition, logic, and reasoning to think about conjectures and theorems and then to obtain the appropriate proofs. Learning mathematics requires many years of schooling and practice to manipulate the different algorithms and to reason according to a whole range of mathematical expertise. Similarly, *reason and logic* are essential to science and to all forms of propositional knowledge, because they tell us about the structure of true propositions and valid inferences. Thus, if astronomy tells us that the earth goes around the sun, this implies that it cannot be true that *it does not.*

One of the problems that is not usually discussed by philosophers is the origin of logic and reason. Logical thinking is *implicitly learned* by dealing with natural phenomena, in the same fashion as we implicitly learn about grammars [5, 6]. Thus, as discussed in Chap. 8, there seems to be no necessity for postulating an abstract metaphysical realm where logic and mathematics reside; logic and mathematics are developed in our brains, where they are physically encoded. Logic and reason do not emerge from mysterious sources. Practical farmers and hunters are logical in their own activities, even if they have never heard about logic. The natural and spontaneous use of logic and reason, which in its most rudimentary form is known as common sense, qualifies as implicitly learned mental skills. All knowledge must be physically encoded by humans, who for a while will continue to be "the measure of all things". What is true in science, in mathematics, and in logic is true in all cultures.

Faith and beliefs are considered by some religious believers to validate knowledge, but they are ineffective in many areas because they cannot be certified as true. In addition, beliefs must be qualified to convey the degree of confidence that they imply. For example, I can say that I strongly believe in a new theory, even when I know that additional tests are needed to confirm it. I may also believe in the use of reason and logic to validate propositional knowledge, because reason and logic have been modeled by mimicking the structure of the real world. Actually, there is much more to validating knowledge, so that all our statements fit without contradictions in a complex body of empirical knowledge and experiences.

10.6 Religious faith cannot be validated

Faith is stronger than belief and is usually restricted to knowing about religion. Faith denotes firmness or certainty as well as commitment and trust. As previously indicated, faith is not a matter of truth, because *faith cannot sense the truthfulness of statements*. Certainty in religion is presumably achieved not only by faith, but also through religious experiences, revelations, and the words of preachers, as well as the interpretation of "sacred" texts. In some cases, it has been reported that faith has been complemented by the voice of God himself, or by that of an angel or a saint. Faith supports doctrines that are expanded and further elaborated by reason, but the truth-value of their assumptions is essentially nil.

Religious beliefs are initially derived from tradition, and are sustained by the cultural environment. This is supported by observation of the geographic and cultural distribution of the different religions. The southern United States Bible Belt, where Protestant fundamentalism is widely practiced, is a clear example of the distribution and propagation of a faith by contiguity, through person-to-person contacts. Today, in the era of telecommunications, the situation is rapidly changing. Religions are spread by televangelists through radio and television shows. For example, according to the New York Times (03/29/06), Mr. Joel Osteen, a modern televangelist with a multimillion dollar operation, has audiences estimated at around seven million viewers a week.

Faith has generated and fueled the plurality of all existing religions, and presumably those that are now extinct. Some religions tend to spread, whereas others disappear. Today, perhaps with the exception of some museum curators and historians, nobody cares about or worships the Gods of the Olympus or the Roman Pantheon. The limited duration of religions is a widespread phenomenon, because there are countless religions that have disappeared, together with the cultures and kings that originated and supported them.

Religious faith is a special relationship between a person and a specific religion. In contrast with reason, logic, or empirical knowledge, faith—as indicated above—does not have the ability to discriminate truth from wishful thinking. Thus, *matters of faith are not relevant to matters of truth*. Faith is a conviction reserved to deal with a presumably supernatural realm, which is otherwise non-demonstrable. However, the structures of these beliefs should be compatible with reason and logic. For example, if "God X" is the "true and only God", then "God Y" is not "the true and only God". Thus, if the initial assumptions of religion X were true, using simple logic, religion X believers could logically demonstrate that *all the other religions are false*. The plurality of contradictory opinions is an indication of their contingency, and presumably of their lack of truth-value.

Looking through a particular glass during the initial part of our life determines in most cases our thoughts and our way of thinking on a specific subject. It is a truism to say that the poet, the preacher, and the scientist see the world differently because they have been conditioned by their education and by their occupation to see it differently. Similarly, all believers think that their own religion is the "true one." Some philosophers and theologians have a more ecumenical idea of religions. As noted by Mortimer J. Adler, Hans Küng believes that "all religions are true" [1]. He seems to confuse truthfulness with the acceptability of a religion for a particular culture. There is a definite new tendency among theologians to think ecumenically about the plurality of religions. Interestingly, even Pope John Paul II [7] became so tolerant to the plurality of religions that he could have been in real trouble if the inquisitors that persecuted Galileo had suddenly been resurrected.

Obviously, religious leaders cannot criticize each other in a deep philosophical fashion, because they would discredit their own way of knowing—faith, the interpretation of the sacred books, or the words of a new prophet. Religious leaders have perceived that, in an age of instant communication and an ever more crowded environment, where the different religious groups are forced to live together, they must tolerate each other, or they will undermine their own faith. The idea that some particular group of people has been chosen by a god generates very few problems in isolated communities. The idea of being chosen is obviously a sign of emotional immaturity, an egocentric idea that is typical of small children and primitive cultures. However, when several "chosen" groups have the divine right to the same territory, the dispute is settled by the sword, with God as a neutral spectator. The tolerance between incompatible religions seems in general to be greater in well-seasoned high priests who can talk to each other than among ordinary people, where the troopers are ready for a fight at the slightest provocation. This pattern of behavior is not exclusive to religion, but is also seen in politics.

Matters of faith are the result of tradition, family beliefs, and personal taste. Thus, the truthfulness of religions cannot be proven, and the fact that religious faith might help people to cope with painful situations does not imply any relationship with truth. Faith in different cultures results in heterogeneous and contradictory beliefs that cast doubts on each other. The problem of the *plurality of religions* has been discussed by Adler [1] in his illuminating book *Truth in Religion—The plurality of religions and the unity of truth.*

The appearance of new religious movements is today a puzzling phenomenon that has been closely studied by many social scientists [8]. Faith is not transcultural and cannot supersede reason and science. Actually, the use of reason to justify religion backfires because reason cannot reconcile how an omnipotent and omniscient God could have created the natural and the moral evils that plague the world today. If our world had been designed and constructed by the Halliburton Corporation, we would certainly have taken them to court and closed their business. The idea of the divine punishment of Adam and Eve is unreasonable and so naïve that it cannot be taken seriously. God not only punished them, but he also *punished all their descendants*, an inconceivable act of cruelty that cannot be attributed to a magnanimous and omniscient god. Few spheres of human activity are so obviously incompatible as religion, science, and justice. Faith is a phylogenetically primitive way to deal with reality. Actually, faith is irrational and cannot be appealed. The problem is that religious believers are convinced that they are in possession of the truth. But it should be evident that having faith in a specific religion contradicts not only other faiths, but also logical and scientific knowledge.

In contrast with faith and beliefs, other choices lie outside of the true–false domain. Such is the case in matters of ethics, esthetics, desirer, fashions, etc. Even if it is true that most of us like chocolate, matters of taste are not universally true or false and are not subject to transcultural truths, although some works of art are almost universally appreciated. These matters are of individual and personal concern, because they cannot be generalized beyond some species-specific preferences. Personal preferences differ from biologically determined desires and experiences that have high survival value and are hard-wired into the brain. However, being beneficial to the species is not a criterion for establishing the correctness of a given behavior.

Truth is often believed to reflect a correspondence between our ideas and objective reality. However, the encoding must be isomorphic, as in propositional knowledge, or have some unequivocal symbolic correspondence. Thus, the transformation between reality and internal symbolism should follow some rules of encoding and decoding that are common to all members of the species. For example, the encoded idea of triangles is isomorphic to all objective triangles, as shown by our capacity to draw them; this does not mean that the neuronal circuits used by different persons must be identical. However, some of what we call true statements are restricted to our own phenomenology, as when we say 'this rose is red'. The statement is correct for us, but Martians with different senses or color-blind humans might think differently, because qualia are contingent to the structure of our brains and senses (see Chap. 5).

10.7 Conclusions

Despite the important role of religion and faith during the early development of societies, the plurality of religious beliefs and their implicit contradictions have never been rationally resolved. Religion and faith have originated the most savage wars and abuses of personal freedom. Theologians and religious leaders have been unable to formulate the criteria necessary to find truth in religious faith, because religious truth is unverifiable and dogmatic. Furthermore, religions are contingent to different cultures, because they are based largely on childhood indoctrination. Subjective conviction in the absence of proof cannot validate beliefs, and beliefs are also disqualified by delusional, mystical, and trance-like experiences. In contrast with religious faith, the empirical truths of science are incompatible with all supernatural beliefs. Science, reason, and logic, working in conjunction, are the only way to validate beliefs free from contradictions and to find the most reasonable hypotheses to explain reality. Indoctrinating children by stressing faith is the most flagrant and perverse violation of human freedom and a sure recipe for future catastrophes.

References

1. Adler JM. Truth in Religion. The plurality of religions and the unity of truth. 1st. ed. New York: Macmillan Publishing Co.; 1990.
2. Tillich P. Dynamics of Faith. 1st. ed. New York: Harper & Row, Publishers; 1957.
3. Fuster JM. The Prefrontal Cortex. Fourth ed. London, UK: Academic Press; 2011.
4. Pierre JM. Faith or Delusion? At the Crossroads of Religion and Psychosis. Journal of Psychiatric Practice 2006;7(3):163–172.
5. Reber AS. Implicit Learning and Tacit Knowledge: An essay on the cognitive unconscious. New York: Oxford University Press; 1993.
6. Inhelder B, Piaget J. The Early Growth of Logic in the Child. Classification and seriation. New York: Humanities Press; 1970.
7. Paul II J. Crossing the Threshold of Hope. First ed. ed. New York: Alfred A. Knopf; 1994
8. New Religious Movements in the 21st Century. Ed. Lucas,P.C.; Robbins,T.

11 Contradictory beliefs are a poor mechanism of adaptation

Summary The acceptance of rationally contradictory beliefs, or schizodoxia,[1] may temporarily postpone intellectual conflicts, social disagreements, and confrontations between political entities, religions or races. It may also serve to maintain a dialogue between different people and different social groups, increasing the likelihood of mutual understanding and respect. Schizodoxia may also help temporarily to maintain the equilibrium of individuals and complex societies. However, contradictory beliefs have disastrous effects when incompatible political systems or religions resort to violence in order to achieve prevalence. In addition, schizodoxic beliefs serve to cover up our animal nature by fostering a variety of religious and supernatural beliefs. Such beliefs provide hope for an afterlife, but negate the scientific evidence and the unity and consistency of truth, which are the only alternatives that can provide understanding and permanent and peaceful solution to human problems.

11.1 Schizodoxia

The simultaneous acceptance of contradictory beliefs or schizodoxia is a compromise that may help us to cope temporarily with situations and doubts which cannot be resolved immediately or which are unpleasant to face. By denying our doubts or by postponing the confrontation of inconsistencies, we can continue to live our lives, even if this requires rationalizing contradictions and looking for auxiliary hypotheses to justify inconsistent beliefs and actions. Unconscious denial is a common defense mechanism that has been described by Sigmund Freud (1856–1939) and followers and plays an important role in

---

[1] From the Greek *schizo*, split, fissure, or crack and *doxa*, opinion. Schizodoxia denotes the acceptance of two incompatible, irreconcilable conceptions of reality. The term to denote an affected individual or attitude would be *schizodoxic*, by analogy with "schizophrenic".

J.M. Musacchio, *Contradictions: Neuroscience and Religion*,
Springer Praxis Books, Popular Science, DOI 10.1007/978-3-642-27198-4_11
© Springer-Verlag Berlin Heidelberg 2012

schizodoxia. Denial of an unpleasant reality is a form of adaptation that can also be observed in everyday behavior, in clinical personality disorders, and in organic diseases of the brain, such as anosognosia (Chap. 4).

The most difficult realization that we all must face is the extinction of the self without the perspective of an afterlife. Bertrand Russell was particularly distressed about the problem, but nobody expressed his desperation more dramatically than Miguel de Unamuno did in his memorable "The Tragic Sense of Life" [1]. He appreciated so much his awareness, life, and manhood that he felt that complete extinction was even more terrifying than spending an eternity in Hell, where his soul could at least continue to exist. The idea of extinction of the self is so intolerable that countless religions have been generated in the hope that there is indeed a better world awaiting righteous individuals after their death. However, all the evidence points to the contrary. Some people thus live with schizodoxic beliefs about the afterlife and try to get reassurance from friends, books, and priests that there really is life after death. However, the *functional* character of the self [2], which replaces the Cartesian *substantial* self, indicates that there is no hope for life after death.

The long-term reliance on schizodoxic beliefs is a fallacy introduced to cope with an unknown future or an unacceptable present, and involving many of the contradictions described in earlier chapters. For example, most religious believers who learn about human evolution in college biology sense a conflict when they realize that science and the teachings of the Bible are incompatible. As mentioned previously, Averroës pioneered the idea of the independent *double truth* to reconcile religion and natural philosophy, so he should be credited with providing the first "reasonable" justification for schizodoxia. Today, the most serious incompatibility between science and religion is made evident through the specific perspective provided by neuroscience and neurology, since these clearly demonstrate that brain diseases destroy specific functions previously attributed to the soul.

As discussed in Chap. 4, loss of memory and the deterioration of the moral sense, abstract capacities, and language occur after vascular damage or localized brain atrophy. The mental alterations produced by brain damage indicate that, even if we assume that the soul exists, it deteriorates during life, so it will not function appropriately after death. Contradictory statements are frequently made in relation to the *mind* and the *brain*, because people believe that the mind represents a *spiritual* component, whereas the brain is necessary only for the soul to express itself. These beliefs contradict the knowledge that the mind is just one of the *functions* of the physical brain. The complexity and the lack of intrinsic unity of the mind are clearly observable when the functions of the brain deteriorate in a piecemeal fashion as described in previous chapters. The natural history of brain diseases directly contradicts the unity and indestructibility of the soul claimed by most religions.

Some people believe that we are fundamentally different from animals. They are right in the sense that our intelligence and moral sense are much more developed than in lower animals. However, the basic biological similarities between humans and animals are shown by the usefulness of rats and mice to test for efficacy, toxicity, and side effects of most of the drugs we use. This clearly indicates that humans and some animals have similar biological mechanisms inherited from a common ancestor, as discussed in Chap. 3. In addition, our capacity to communicate effectively with domestic animals, using gestures and an elementary language, suggests that, if we have a soul, then dogs, bonobos, and chimpanzees must also have a soul. In fact, many dog owners believe that this is the case. The animal sacrifices described in the Bible therefore suggest that early religious practices were cruel, and that all those who justify such practices should be considered not only schizodoxic but also barbaric. As a matter of fact, some religions and societies have not abandoned their primitive behavior, as indicated by condemnations to death by public stoning of women accused of adultery, as recently as August 2010. What is most worrisome is that these executions are supported not only by the civil and religious authorities, but also by the public and even by the families of the sinners. Under Shariah law, stoning, amputations, and lashing are considered appropriate forms of punishment and are still commonplace.

We have frequently read statements that attribute major disasters caused by wind, floods, and earthquakes to "acts of God". At the same time, and in a schizodoxic fashion, these reports may attribute the survival of a single believer, miraculously protected against all odds by a beam or a strong wall, to God's intervention. The survivors usually appear on television and thank God for protecting their life, but this seems to imply that the believers who died in the same catastrophe were not protected by God. Is it possible that God decided to save some believers but not other believers from the same family?

According to the Old Testament, Sodom and Gomorrah were destroyed because of their wickedness and depravity, but it is hard to understand why innocent children were also killed in the catastrophe. Today, the tune has changed, and God is less often blamed for catastrophes, probably because religious authorities have realized that making God responsible for tornados, hurricanes, and earthquakes, which kill people indiscriminately, is not good for God's public image. However, after Hurricane Katrina killed more than 1,800 people and partially destroyed New Orleans, LA, and several adjacent areas in August 2005, some discontented preachers of unsuccessful churches suggested that Katrina was God's *punishment* for the sinful lifestyle of the region. Thus, the tendency to believe that earthquakes, hurricanes, and other natural disasters reflect God's wrath is still persistent in today's world, despite the wealth of scientific evidence that earthquakes, tsunamis, and storms are natural catastrophes.

11.2 Schizodoxia contributes to religious tolerance and maintenance of complex societies

The various conditions found in the natural and cultural world have created an almost infinite number of religions, as well as political and economic systems that oppose each other. This diversity has increased the varieties of schizodoxia, each of which represents a different form of adaptation. One of the most positive consequences of schizodoxia is that, after a few thousand years, some religious leaders and politicians are learning to tolerate each other, because there is no other reasonable alternative besides becoming atheist or agnostic. This is particularly true of today's evolved religions in Europe, although not necessarily on the American continent, where presidential candidates argue about "Who is most religious of all?"

The increase in political and religious tolerance serves to maintain a dialog between groups which may eventually increase the chances of mutual understanding and respect. However, the increase in religious tolerance is not universal, with the Middle East being a notable exception. In this region, the hard-core orthodox groups associated with ethnic backgrounds and national boundaries are continuously involved in religious wars and persecution. The most deadly combination is the association between a religion, a proud ethnicity, and the desire to expand national frontiers. This potentially explosive combination is exaggerated in densely populated areas with high natality. Hopefully, this situation will be at least partially resolved in the near future, even though the increase in respect that some religions now have for others is largely a strategy of self-defense. Those with the greatest tolerance have learned that shifts in the political, economic, or arms balance could change the support for a different denomination and start a new cycle of revenge.

The prevalence of many different forms of schizodoxia in today's world largely reflects its usefulness in temporarily maintaining stability and the orderly existence of complex societies, even if their relationships are less than perfect. Politicians and the dominant minorities use schizodoxia to manipulate society by promising a better life for everybody. Political and religious leaders are the purveyors of collective purposes that control the world in which we live. Politicians, priests, and the dominant classes blame the large corporations for our economic crisis and unemployment while shifting the burden of personal sacrifice and economic responsibility to the uneducated majority and illegal immigrants, who must take the least desirable jobs and hope that their individual souls will be rewarded in the afterlife.

The cult of heroes, such as soldiers, firefighters, and police officers, is designed to compensate for personal hardship or death in the service of society, country, or religion. This is obviously not a recent political invention, because the Persians, Greeks, and Romans used to provide the highest honors for soldiers fallen in the name of God, society, or country. Today in America, the families of victims of unjustifiable foreign wars are comforted by public gratitude, medals, and ceremonial burials at Arlington National Cemetery. Moreover, all the victims are publicly recognized as valuable elements of our society. However, no monetary compensation, public ceremonies, or letter from the President can replace a lost member of the family. The creation of a voluntary army in America was a Machiavellian maneuver to ensure that the children of prominent politicians and rich families never have to comply with military service or go to war. Our country now has permanent professional armed forces that make compulsory enrolment unnecessary. Many individuals enroll voluntarily because it is the only way they have found to obtain an honorable and acceptable socio–economic status for themselves and their families.

Despite the contribution of schizodoxia in maintaining temporarily complex societies, the practice of schizodoxia by extremists allows them to make outrageous and frankly delusional statements that defy reason, common sense, and simple compassion, all in the name of some supernatural power. What I have in mind is a group of extremists led by Fred Phelps, the founder of Westboro Baptist Church in Topeka, Kansas, who have disrupted several funerals of soldiers, whether or not they were gay, proclaiming that *God is killing Americans in Iraq* to punish us for condoning homosexuality. The grieving soldier's relatives have been suing Mr. Phelps for disrupting the soldier's funerals with signs that say, "Thank God for dead soldiers", "God hate fags", etc. The case was presented to the Supreme Court on October 6, 2010 [3]. Thus, even though our complex society may temporarily function better by accepting both political and religious deceptions, such contradictions will no doubt generate many future social and economic conflicts. Schizodoxia never provides long-term solutions.

11.3 Accepting the benefits, but not the implications of science

Many schizodoxics, including more than 80% of Americans, deny that we have evolved from more primitive forms of life, while they accept new drugs and medical treatments that were discovered by testing them on rats, mice, and dogs. Thus, they accept the benefits of animal experimentation, but not the implications that we have fundamental biochemical and physiological similarities. We should remember that insulin was discovered by F. Banting and C. Best working with dogs in the early 1920s. Since then, most hormones and body constituents have been discovered using animals to investigate human diseases. The benefits from science and medicine are easily acceptable by schizodoxics, but the implication that we have evolved from common animal ancestors is not acceptable to them, because it contradicts the core of their religious and supernatural beliefs. Even if Linnaeus proposed to call us *Homo sapiens*, we can easily become extremely savage animals, and torture and then kill members of our own species. Thus, there is no evidence that we necessarily deserve the appellation *sapiens*.

The simultaneous acceptance of religious dogma and the benefits provided by science and medicine are not malicious, because they are based primarily on ignorance. The prevalence of this largely benign form of schizodoxia means that the majority of people have gained popular respect for medicine. They do not realize that this has required scientific understanding of nature, which is incompatible with their religious beliefs. Their beliefs are based on oral stories recounted by ignorant peasants and collated in the Bible and other books several centuries ago. These discrepancies, as indicated in Chap. 3, first originated when the Vatican tried to silence the voice of Galileo Galilei, who was persecuted and imprisoned by the Inquisition for proposing the heliocentric theory of the solar planetary system which contradicted the Scriptures. The empirical findings, however, could not be ignored for long, so the religious leaders, with the help of Averroës, were forced to postulate the existence of two independent metaphysical realms, the *natural* and the *supernatural*. Today, as indicated above, this form of schizodoxia affects more than 80% of the American population.

11.4 Supernatural explanations are not logically acceptable

Supernatural entities are not acceptable as scientific explanations or as agents of change, because neither their existence nor their causal efficacy can be demonstrated. Today, we can scientifically explain diseases and epidemics that used to be the exclusive domain of magic or religious beliefs, in which devils and witches were involved. Processes previously considered supernatural lost their mysterious character when explained by science and reason. In contrast to religious dogma, scientific explanations are not always final pronouncements, but they are seeds that will eventually result in medical and technological progress. For example, genetics not only validated Darwin's principles of evolution, but also provided the initial methods for establishing lineages of humans, animals, and plants. Moreover, it allows us to diagnose and prevent certain hereditary diseases. Genetic manipulations are a key to modern approaches to disease prevention and manufacture of vaccines and drugs, and in a not so distant future, they will probably be used to supply organs or tissues for surgical replacement. Moreover, knowledge of inheritance mechanisms has established the basis for biotechnology and agricultural techniques that are increasing the yields of many crops and making possible the survival of entire human populations that were otherwise doomed to starvation. Science and medicine are changing the face of the earth. In the future, evolution will be guided by *intelligent scientific interventions*.

Scientific explanations are now taking the place of supernatural explanations. However, the shift from supernatural to natural explanations has created several transitional problems. Those that believe simultaneously in Darwinian evolution and in the Bible are schizodoxic and confused. This has forced the religious establishment to adopt two different strategies to deal with the incompatibility of religion and science. The first strategy is to try to teach *intelligent design* in schools. Fortunately, the teaching of camouflaged religions at state-run schools has been rejected by American courts, because it is against the American Constitution. The second approach consists in accepting the coexistence of science and religions, as if they were compatible. This schizodoxic strategy comes in several flavors, with two extreme variants that go from peaceful coexistence to complete disregard for what science has to say about human nature.

Some religious scientists have proposed that science and religion belong to different metaphysical realms, which can coexist peacefully, side by side. These believers do not understand, or prefer not to think about, the fact that religious dogmas are incompatible not only with the truths of science, but also with the dogmas of other religions. The incompatibility between several religions is the driving force behind some of the bloodiest conflicts in the world today.

The Anglicans in England are so arrogant that they will not even discuss the possibility of considering science as an opposing force. However, when they get sick, they will very likely turn to science and medicine for help. Otherwise, they might follow the fate of the Christian Scientists and the teachings of Mary Baker Eddy, which are fading as a religion [4]. There have been many unnecessary deaths among the children of Christian Scientists, because their parents relied exclusively on praying, instead of allowing them to receive appropriate medical treatment.

The impact of science and medicine on our lives and on sexual freedom during the last 50 years cannot be ignored. Some have therefore suggested that religious dogma should be modified to suit the new reality and personal taste. The most difficult issues are those related to contraception, abortion, and sexual preferences. Catholics in the USA have quietly rebelled against the Church regarding contraception, which some Americans practice without inhibitions. This indicates that they do not take the details of their religion seriously, or that they do not understand that religious dogma cannot be changed to satisfy personal needs or the fashion of the times. However, groups of conservative fundamentalists and militant Christians still oppose early-term abortions, sometimes violently, because they believe that the human soul is created and embodied by God in every egg when conception occurs. Some militant Christians have not hesitated to kill doctors who perform abortions because they think that they are actually killing "babies".

Another area of disagreement is that, after fierce opposition, some religions are starting to accept homosexuality. Sexual preferences are most likely the result of genetic [5] and cultural determinants, which cannot be modified. However, most religions, in their original forms, object to homosexuality, which they consider a perversion. This seems reasonable because, if sex without intention to procreate is a sin, most heterosexual and all homosexual relations would then be sinful in the eyes of the Christian church. Some Anglican denominations are becoming more tolerant in the acceptance of homosexual priests, but in doing so, they fail to recognize that they do not have the authority to change the word of their own God. There has been some progress, however, because as recently as December 2010, the US Congress repealed the law that kept recognized homosexuals out of the Armed Forces.

All the discrepancies between science and religion are a reflection of the classic incompatibility between science and reason on the one hand and faith on the other. As previously discussed, people have accepted faith as a road to revelation that may result in the uncontrollable proliferation of many incompatible religions. Paradoxically, no religion can be demonstrated to be *the only true and exclusive word of God*. The increased respect for other people's rights and the efforts and good intentions of many religious leaders to coexist with each other and to unite against science will eventually generate countless varieties of extremely tolerant religious denominations. Hopefully, these will in their turn slowly weaken and dissolve into oblivion.

11.5 The attribution of biological evolution to god is a sophisticated form of schizodoxia

It is worth repeating here that there is substantial undeniable evidence for the animal origins of humans. According to archeologists, the *Homo* and chimpanzee-bonobo ancestors separated 7 million years ago (Mya), but our genetic differences are surprisingly small, considering that we are still 96% identical to chimpanzees at the DNA level. The fossil record also indicates that human evolution was a gradual and progressive ascension of an imaginary tree, in which several branches have since disappeared. However, there are many missing links and several dead branches. The first species found in a state that could be identifiable with certainty as a human precursor is genus *Ardipithecus*, species *ramidus* or "Ardi", who was found in Ethiopia and lived about 4.4 Mya, as discussed in Chap. 3; she was bipedal, about 47 in. tall, and weighed about 110 lb [6–8]. The next series of well-preserved hominids were of the genus *Australopithecus*, who were well represented by *Australopithecus afarensis*, which includes Lucy and lived three Mya. Our ancestors of the genus *Homo* appeared roughly about 2.3 Mya with *Homo habilis*, and continued to evolve into different branches and species, such as *Homo erectus*, *Homo heidelbergensis*, *Homo neanderthalensis*, and *Homo floresiensis*. The last species of the genus *Homo* is *Homo sapiens*, which is relatively young, being only about 30,000 years old [9–11].

Since Darwin, the vast majority of scientists have been convinced that evolution has been driven by natural causes without any supernatural intervention. Actually, there are no indications that there ever has been any supernatural process. One of the few outstanding scientists known for not believing in natural evolution is Dr. Francis S. Collins, whom President Obama appointed as Director of the National Institute of Health. Dr. Collins believes that evolution was actually the work of God, and he wrote a very ingenious and entertaining book in which he attributed the human changes in DNA to the hand of God [12], even though he does not present any evidence for God's existence. It would have been more reassuring to have solid and unmistakable evidence that we were created by a God who gave us an immortal soul and eternal life, as described in the Scriptures. Echoing what Miguel de Unamuno said, I would not mind having a soul.

With the exception of Collins, few prominent biologists, if any, believe that evolution was actually conceived and guided by God. It is obvious that an omnipotent and omniscient God could have created humans without first experimenting with so many hominids that suffered, failed, and most likely had a miserable and violent death. He could have created a world without so much unnecessary fighting and suffering. The survival of the fittest is a cruel ordeal that made wars an integral part of our life.

In any terrestrial court of law, God would be found negligent on many counts that would not only include the invention of cancer and obnoxious bacterial and degenerative diseases, but also the creation of the Devil, if there were any real evidence for the latter. Religious lawyers might argue that diseases and obstacles, like the many plagues of Job, were necessary to test the strength of man's religious convictions. As a matter of fact, a god should not need to test the perfection of his own creation. Despite all religious beliefs, however, there is no scientific evidence of the existence of God, guardian angels, or wicked devils and witches. The introduction of good and bad supernatural forces in daily life would negate causality, true knowledge, and science. In consequence, the whole universe would already be chaotic and unpredictable by logic and reason.

Even if our ignorance is in principle incommensurable, we know that we can gradually reduce it by concentrating on scientifically *testable hypotheses*. Life and the universe are becoming gradually less mysterious and more understandable through science and space exploration. Actually, we no longer require any supernatural force to explain life, death, or the universe. And future generations may well enjoy the pleasures of having a larger, genetically engineered brain.

11.6 The grim consequences of certain forms of schizodoxia

Despite some of the short-term benefits of schizodoxia, there are circumstances in which it could have catastrophic consequences, not only at the international, but also at the interpersonal level. The deceptive and malicious schizodoxia that we see in many politicians and promotional advertisers ultimately fails, leaving behind it a long trail of unhappiness. Some political or economic campaigns are designed to convince gullible individuals to accept or vote for promises that are too good to be true. Other forms of schizodoxia are pathological or self-destructive, because they seem to be specifically designed to backfire. This is exemplified by what President Richard M. Nixon did in 1973. Nixon installed a secret tape system that recorded many of his conversations in which he made derogatory remarks about Jews, blacks, Italian–Americans, and Irish–Americans. In addition, he provided evidence of his involvement in the Watergate scandal that was triggered by his authorization to enter illegally into the Democratic National Committee Headquarters, located in the Watergate Building complex, to steal confidential information. The tapes that directly implicated Nixon are now in the Nixon Library, which recently released some of their content [13].

11.7 Schizodoxia also negates the consistency of truth

Truth in science and philosophy is understood as a series of statements that symbolize or faithfully describe facts through verbal or mathematical descriptions. Statements are true when they encode the natural world and symbolize its structure and dynamics. The consistency of truth is an ideal that cannot be challenged, because it is determined by the consistency of Nature. The infinite variety of natural processes, which we perceive in many different forms, may *oppose* each other, but they are not *ontologically* contradictory in the way that opinions could be. There is no reason to believe that we will ever find true ontological contradictions, because this would mean finding something that is *simultaneously* P and *non*-P. Thus, we can only doubt the coherence of human beliefs and statements, but not that of Nature.

The contradictions between scientific statements that we sometimes find in the literature are derived from incomplete knowledge, which always triggers a more critical search for a better explanation, and invariably results in the progress of science and knowledge. The discovery of new facts changes our detailed view of nature, but it does not create ontological contradictions. If we rely on science, philosophy, or common sense, we cannot accept that nature could be either contradictory or controlled by non-natural or *super*-natural forces.

A set of statements about the same fact must be consistent, but consistency is not in itself a criterion of truthfulness, because an isolated set of statements could be false in its entirety. However, any lack of consistency indicates that at least one of the statements is incomplete or false. Facts can be encoded through language, experiences, or both, but truth requires verbal statements to agree, or be *symbolically isomorphic* with reality. Apparent contradictions are actually good catalysts for additional research.

The consistency of truth can be defined as the lack of contradictions between true statements or between statements and reality, whereas isomorphism refers to the symbolic correspondence between reality and the language-independent brain encoding of neural models or surrogates of the same fact (Chap. 7). For example, triangles and other objects are *encoded* isomorphically in the brain, because they can be internally visualized and faithfully drawn on paper, even if we cannot detect visible triangles by dissecting a brain. We conclude, for many reasons, that most knowledge is encoded or represented in our brains as neural surrogates in neuronal circuits, even if we do not know exactly how the process takes place. We perceive the equality or inequality of small numbers of objects by subitation. For example, $1 + 1 = 2$, even if we do not know how the encoding takes place in our brains. This is particularly true for the representation of small numbers, which in its origins was *digital* or one to one (see Fig. 8.1). However, we cannot perceive the actual structure of the brain encoding, which is a language-independent dynamic process.

Theories of truth must refer to important, nontrivial, and qualia-independent subsets of observations. Thus, isomorphism, correspondence, and consistency (lack of contradictions) are essential components to determine the truthfulness of thoughts, statements, and theories. Empirical verification and the predictability of their logical consequences are also essential characteristics. Even if most of us have accepted some contradictory beliefs early on in our cultural development, physicists and philosophers agree that there should be no contradictions between natural laws, principles, and reality. And we know from scientific observations and common sense that there is consistency in what we call universal laws, and in the principles that we use to explain the universe.

References

1. de Unamuno M. The Tragic Sense of Life in Men and Nations. First Princeton Paperback ed. Princeton: Princeton University Press; 1977.
2. Frondizi R. The Nature of the Self: A Functional Interpretation. Arcturus Book Edition ed. Carbondale and Edwardsville: Southern Illinois University Press; 1971.
3. Gregory S. Time 146[14], 31–34. 2010. New York.
4. Cather W, Milmine G. The Life of Mary Baker G. Eddy and the History of Christian Science. 1st ed. Lincoln and London: University of Nebraska Press; 1993.
5. LeVay S. The Sexual Brain. 1st. ed. Cambridge, MA: The MIT Press; 1993.
6. Lovejoy C.O. Reexamining Human Origins in Light of Ardipithecus ramidus. Science 326, 74e1–74e8. 2009.
7. White TD, Asfaw B, Beyene Y et al. Ardipithecus ramidus and the paleobiology of early hominids. Science 326, 75–86. 2009.
8. Lovejoy.C.O., Suwa G, Simpson SW, Matternes JH, White WF. The great divides: Ardipithecus ramidus reveals the postcrania of our last common ancestors with African apes. Science 326, 100–106. 2009.
9. Balter M. Paleoanthropology. New work may complicate history of Neanderthals. Science 326, 224–225. 2009.

10. Walker A, Stringer C. The first four million years of human evolution. Philos Trans R Soc Lond B Biol Sci 2010;365:3265–3266.
11. Stringer C. Modern human origins: Progress and prospects. Philos.Trans.R Soc.Lond B Biol.Sci. 357, 563–579. 2002.
12. Collins FS. The Language of God. A Scientist Presents Evidence for Beliefs. First ed. ed. New York, London, Toronto, Sydney: Free Press; 2006.
13. Nagourney A. In tapes, Nixon rails about Jews and Blacks. The New York Times 2010 Dec 11;A13.

Index

A

Abacus, 95, 130
Aboutness
    innate, 74, 75
    phenomenal, 74, 75
Abstract capacity, 7, 139
Abstraction, 55, 56, 114, 123, 125, 128–130,
        133–140, 144, 155
    neural bases, 138–139
Abstract objects, 122–124, 126, 135, 137–140
    metaphysical, 123, 124, 126, 135, 138
Abstract symbols, 83
Alien hand syndrome, 49
Alzheimer's disease, 47, 54–55, 101, 105,
        106, 139
America, 9, 180, 181
Americans, 15, 17, 121, 181, 183, 186
Amygdala, 71, 106
Anamnesis, 144
Ancestors, 1, 2, 5, 6, 13, 16, 27–35, 59, 86,
        158, 167, 179, 181, 184
Anosognosia, 52, 178
Anthropoids, 27, 88, 100
Anton syndrome, 52
Anxiety, 10, 70, 76, 90
Aphasia, 93, 137
Aplysia, 101, 111
Apprentice, 112
Aristotle, 21, 67, 100, 144–147, 156
Arithmetic
    empirical roots, 127
    foundation, 125, 126, 146
Arithmetical propositions, 125, 126
Arithmetic system, 97
Astrology, 16
Attention, 10, 25, 54, 65, 66
Attitude, 49, 76, 86, 90, 125, 138, 168, 177
Aversive stimulation, 101

B

Axioms, 97, 132–134, 144, 145, 148, 150,
        151, 154, 156
    self-evident, 151

B

Basal ganglia, 106
Beliefs
    ancestral, 18
    Bible, 182, 183
    supernatural, 5, 6, 27–30, 170, 181
Biblical explanations, 11, 24
Binding, 62, 63, 71–74, 77
Biological evolution, 9, 15, 89
    attribution to God, 184–185
Blind spot, 51
Body, 3, 4, 7, 8, 10, 11, 16, 18, 19, 21,
        28, 39–45, 55, 56, 61, 65, 69, 95,
        109, 112, 114, 122, 132, 133,
        172, 181
    image, 47–49, 52
    physical nature, 60
    scheme, 77
Brain
    circuits, 104, 112, 157, 159
        analogue, 157
    damage, 178
    electrical stimulation, 40, 45, 46, 65, 105
    lesions, 10, 40, 43, 47–49, 51, 52, 54, 78,
        104, 136, 157
    mechanisms, 44, 64, 65, 78
    networks, 48
    structures
        homology, 90, 96
        uniformity, 90, 96
    surgery, 45, 55, 104
    transparency, 60
Brave New World, 15

Printed by Publishers' Graphics LLC